MICROBIAL BIOREMEDIATION

MICROBIAL BIOREMEDIATION

P RAJENDRAN

Reader
Department of Zoology
Vivekananda College

Madurai

P GUNASEKARAN

Professor
Centre for excellence in Functional Genomics
Madurai Kamaraj University
Madurai

MJP PUBLISHERS
Chennai 600 005

Copyright © Publishers, 2024
All rights reserved
Printed and bound in India

MJP PUBLISHERS
A unit of Tamilnadu Book House
47, Nallathambi Street
Triplicane, Chennai 600 005

PREFACE

Nature took 600 million years to develop the main components of the environment—atmosphere, hydrosphere and troposphere—that exist today. But the existence of man for barely one million years has brought about changes that threaten the very existence of the healthy atmosphere. Population explosion and urbanization with global industrialization have placed a major pressure on our environment, potentially threatening its susceptibility. Over a period, several anthropogenic and xenobiotic chemicals are introduced into the natural ecosystems. This has resulted in the build-up of chemical and biological contaminants having detrimental effects on living beings throughout the biosphere. At present, the humanity is at the crossroads to find solutions to environmental problems created by local, national and global developmental activities. There has been a worldwide realization on the need to be aware of, and knowledgeable about the environment, its benefits and the measures to mitigate the contaminants. Research and technology development and implementation to address waste issues now rank high in the world. The current conventional and commercial waste treatment strategies such as contaminant excavation and disposal are inefficient and costly, and lead to the formation of toxic intermediates. The need of the hour is to develop an innovative approach to solve this problem.

Bioremediation which is the use of living organisms such as plants (Phytoremediation) and microbes such as bacteria, algae and fungi (Microbial bioremediation) or their systems to treat the contaminants, is an efficient, eco-friendly and economical novel alternative to conventional treatment technologies. Removal of contaminants from the environment using microbes often takes place without human intervention (intrinsic bioremediation) and without causing any deleterious effects to the pristine nature of the environment. Hence, the current technology is imperative to rectify most of the shortcomings of the conventional treatment technologies.

This book comprises eight chapters, dealing with various aspects of bioremediation in detail with precise diagrammatic representations, which will help the reader to appreciate the concepts without impediments. The introductory chapter analyses the current environmental tribulations and the global concern. The second chapter categorizes different environmental contaminants, their sources and effects on living organisms. The third chapter deals with the process of bioremediation, mechanisms, types, success stories, monitoring strategies and the feasibility of utilizing the technique in detail. The fourth chapter explains the various aspects of biodegradation and its role in the treatment of organic wastes. The fifth chapter provides information about classification and treatment techniques for organic, liquid and solid wastes. It also deals with the microbial degradation mechanisms of lignin, cellulose and pectin. The sixth chapter highlights the degradation of specific contaminants such as petroleum hydrocarbons, crude oil, acid mine drainage, BTEX, phenol, PCBs and dyes. The seventh chapter deals with the bioconversion of non-biodegradable specific pollutants such as heavy metals, pesticides, polymers, plastics, dioxins and radioactive wastes. The eighth chapter enlightens the reader on the genetics of microbial degradation and the possibilities of modifying the genetic make-up of the microbes (GMOs) for bioremediation purposes. The Glossary section provides a detailed account of the frequently used bioremediation terminology. An extensive list of references and websites related to bioremediation is also provided.

This book is intended to inculcate the present status, feasibility, and the significance of microbial bioremediation to the academicians, students, researchers, environmentalists, agriculturalists, industrialists, professionals, engineers and the public who are enthusiastic to conserve the Mother Nature.

P. Rajendran

P. Gunasekaran

CONTENTS

FIGURES

TABLES

ABBREVIATIONS

ACGIH	American Conference of Governmental Industrial Hygienists
ACL	Alternate Concentration Level
ADU	Acceptable daily Uptake
ADME	Adsorption, Distribution, Metabolism and Excretion
ASP	Activated sludge process
ASTM	American Society for Testing Materials
atm	Atmosphere
ATSDR	Agency for Toxic Substances Disease Registry
BACM	Best Available Control Measures
BAT	Best Available Technology
BCF	Bio Concentration Factor
BCF	Bio Accumulation Factor
BCT	Best Conventional Technology
BEI	Biological Exposure Index/Indices
BOD	Biological Oxygen Demand
BP	Boiling point
BSC	Board of Scientific Counsellors
BTEX	Benzene, Toluene, Ethylbenzene and Xylene
Btu	British thermal unit (a unit of energy)
BWR	Boiling water reactor
CAA	Clean Air Act
CAFÉ	Clean Air for Europe
CDC	Control for Disease Control
CDM	Clean Development Mechanism
CEC	Commission for Environmental Cooperation
CEL	Cancer Effect Level
CERCLA	Comprehensive Environmental Response, Comprehension and Liability Act
CFC	Chloroflurocarbon
CFR	Code of Federal Regulations

Ci	Curie
CLP	Contract Laboratory Program
cm	Centimetre
CMC	Commercial Metals Company
COD	Chemical Oxygen Demand
CPSD	Consumer Product Safety Division
CSO	Combined Sewer Overflow
CWA	Clean Water Act
DBP	Disinfection by-product
DCE	Division of Chemistry and Environment
DDE	Dichlorodiphenyldichloroethlene (DDT degradation product)
DDT	Dichlorodiphenyltrichloroethane
DEHP	Di (2-ethylhexyl) phthalate
DES	Diethylstilbestrol
DfE	Design for the Environment
DHEW	Department of Health, Education and Welfare
DHHS	Department of Human Health and Services
Dioxin	2,3,7,8-TCDD
DMSO	Dimethysulphoxide
DO	Dissolved oxygen
DOE	Department of Energy
E	Toxicity (RCRA)
ED_{50}	Median Effective Dose
EHS	Extremely Hazardous Substances
EIA	Energy Information Administration
EIS	Environmental Impact Statement
EPA	Environmental Protection Agency
EPR	Extended Producer Responsibility
ETS	European Tobacco Smoke
FAD	Flavin Adenine Dinucleotide
FAO	Food and Agricultural Organization
FDA	Food and Drug Administration
FFV	Flexibility fuelled vehicle
FIFRA	Federal Insecticide, Fungicide and Rodenticide Act
FMN	Flavin Mononucleotide
FQPA	Food Quality Protection Act

FR	Federal Register
g	gram
GBT	Glycol biotreatment
GC	Gas Chromatography
GCM	General Circulation Model
GEM	Genetically Engineered Microbes
GEO	Genetically Engineered Organism
GM	Genetically modified
H	Actually hazardous (RCRA)
HAB	Harmful algal bloom
HAP	Hazardous Air Pollutant
HAZMAT	Hazardous Material
HCFC	Hydrochloroflurocarbon
HDPE	High-density polyethylene
HFC	Hydroflurocarbon
HHW	Household Hazardous Waste
HPLC	High Performance Liquid Chromatography
I	Ignitability (RCRA)
IARC	International Agency for Research on Cancer
ICR	Ignitable, Corrosive, Reactive
IDLH	Immediately Dangerous to Life and Health
INDOEX	Indian Ocean Experiment
ISO	International Organization for Standardization
kd	adsorption ratio
kg	kilogram
kkg	metric tonne
k_{oc}	organic carbon partition coefficient
k_{ow}	octanol–water partition coefficient
L	Litre
LC	Liquid chromatography
LC_{50}	Lethal concentration, 50% kill
LCA	Life-cycle Assessment
LC_{lo}	Lethal concentration, low
LD_{lo}	Lethal Dose, low
LDPE	Low-density polyethylene
LDR	Land Disposal Restriction

LiP	Lignin peroxidase
LOAEL	Lowest-observed-adverse-effect
LSE	Levels of Significant Exposure
m	metre
MACT	Maximum available control technology
MCL	Maximum Contaminant Level
MCLG	Maximum Contaminant Level Goal
mCurie	millicurie
MEI	Maximally Exposed Individual
mg	milligram
MIC	Methyl isocyanate
mL	millilitre
mm Hg	millimetres of mercury
mm	millimetre
mmol	millimole
MnP	Manganese peroxidase
mo	month
MOPITT	Measurements of Pollution in the Troposphere
MPN	Most Probable Number
mppcf	millions of particles per cubic feet
MRL	Minimum Risk Level
MS	Mass Spectrometry
MSW	Municipal Solid Waste
MTBE	Methyl tertiary Butyl Ether
MTD	Maximum Tolerated Dose
NADH	Nicotinamide Adenine Dinucleotide
NADPH	Nicotinamide Adenine Dinucleotide Phosphate
NAPL	Non-Aqueous Phase Liquids
NAS	National Academy of Sciences
ng	nanogram
NHANES	National Health and Nutrition Examination Survey
NIEHS	National Institute of Environmental Health Sciences
NIOSH	National Institute for Occupational Safety and Health
NIOSHTIC	NIOSH's Computerized Information Retrieval System
nm	nanometre
nmol	nanomole

NOAEL	No Observed Adverse Effect Level
NOC	National Occupational Classification
NOES	National Occupational Exposure Survey
NPL	National Priorities List
NPR	Natural Planning and Response System
NRC	National Research Council
NTIS	National Technical Information Service
NTP	National Toxicological Program
ODP	Ozone-depletion Potential
OPA	Oil Pollution Act
OSHA	Occupational Safety and Health Administration
p^2	pollution prevention
PAH	Polycyclic aromatic hydrocarbon
PAN	Peroxyacetyl nitrate
PBDE	Polybrominated diphenyl ether
PBT	Persistent Bioaccumulative and Toxic
PBT	Persistent, bioaccumulative, toxic
PCA	Pollution Control Agency
PCB	Polychlorinated biphenyl
PCDD	Polychlorinated dibenzodioxins
PCDF	Polychlorinateddibenzofurans
PEL	Permissible Exposure Limit
PERC	Perchloroethylene
PET	Polyethylene tetraphthalate
PFC	Perfluorocarbon
PFOS	Perfluorooctane sulphonates
pg	picogram
PHA	Polyhydroxyalkonate
PHB	Poly(3-hydroxybutrate)
PHBV	Polyhydroxybutratevalerate
PHR	Products hazard review
PHS	Public Health Service
pmol	picomole
PMR	Proportionate Mortality Rate
PNA	Polynuclear Aromatics
Po	Polanium

POM	Particulate Organic Matter
POP	Persistent Organic Pollutants
ppb	parts per billion
ppm	parts per million
ppt	parts per trillion
PVC	Polyvinylchloride
PWR	Pressurized Water Reactor
QA	Quality Assurance
R	Reactivity
RCRA	Resource Conservation and Recovery Act
RCRALDR	Resource Conservation and Recovery act Land Disposal Restrictions
REL	Recommended Exposure Limit
Rfd	Reference dose
Rn	Radon
RTB	Retention Treatment Basin
RTECS	Registry of Toxic Effects of Chemical Substances
SDWA	Safe Drinking Water Act
SE_6	Sulphurhexafluoride
sec	seconds
SIC	Standard Industrial Classification
SMR	Standard Mortality Ratio
SPIC	Southern Petrochemical Industrial Corporation
SRB	Society for Reproductive Biology
STORET	Storage and Retrival
STEL	Short-term Exposure Limit
T	Toxic (RCRA)
TBT	Tributyilin
TCDD	2,3,7,8-tetrachlorodibenzo-*p*-dioxin
TCE	Repetition of item No4
TCE	Trichloroethelene
ThOD	Theoretical Oxygen Demand
TLV	Threshold Limit Value
TNT	2,4,6-trinitrotoluene
TRI	Toxic Inventory Release
TRI	Toxic Release Inventory
TSCA	Toxic Substance Control ACT

TUR	Toxics Use Reduction
TWA	Time- Weighted Average
UASB	Upward-flow Anaerobic Sludge Blanket
UBH	Urea Biohydrolysis
UEL	Upper Explosive Limit
UF	Uncertainty Factor
UNCED	United Nations Conference on Environment and Development
UNEP	United Nations Environmental Program
USGS	US Geological survey
VOC_s	Volatile Organic Compounds
VSD	Virtually Safe Dose
WAP	Waste Analysis Plan
WCED	World Commission on Environment and Development
WPEL	Work Place Exposure Limit
WHO	World Health Organization
Wk	week
WMH	Waste Management Hierarchy
WWTF	Waste-water Treatment Facility
yr	year
ZEV	Zero-emission Vehicle

MILESTONES IN THE HISTORY OF ENVIRONMENTAL AGREEMENTS AND SUSTAINABLE DEVELOPMENT

1962   Rachel Carson's *Silent spring*

1969   Organization of African Unity

1972   *The ecologist*

1972   The first UN Conference on the Human Environment

1980   World Conservation Strategy

1982   The second UN conference on the Human Environment at Nairobi, Kenya

1987   World Commission on Environment and Development (WCED)

1987   The Montreal Protocol

1992   The third UN conference on Environment and Development at Rio De Janeiro, Brazil, Earth Summit, Rio 92-Agenda 21.

1994   UN Conference on Population and Development, Cairo, Egypt.

1997   Kyoto Protocol, Dec. 11, Japan.

1997   Rio +5, March 13–19, Rio de Janeiro, Brazil

2002   Rio +10, World Summit on Sustainable Development or Johannesburg Summit, Aug. 26–Sep. 4.

INTRODUCTION

1

GLOBAL AND INDIAN ENVIRONMENTAL CRISIS

The United Nations Environment Programme (UNEP)'s comprehensive global state of environment report called the *Global Environment Outlook-3 (GEO-3)*, the *State of the World-2004* report of the Worldwatch Institute and the report of The Centre for Science and Environment (CSE), New Delhi, the *Citizen's Fifth Report*, indicate the current global and Indian environmental crisis situation.

Pollution

* At least one billion people around the world breathe unhealthy air, and three million die annually from the effects of air pollution.

* The World Health Organization (WHO) consistently rates New Delhi and Kolkata as being among the most polluted megacities in the world.

Global warming

* The period September 2003 to November 2003 was the warmest quarter in recorded history. The permafrost and glaciers in the polar and other regions are melting, and the resultant rise in the sea level is threatening many small islands.

* In May 2002, the summer temperatures in Andhra Pradesh rose to 49°C, resulting in the highest one-week death toll on record. Glaciers in the Himalayas are retreating at an average rate of 15 m per year.

Urbanization

* About half the world's population now lives in urban areas, as compared to little more than one-third in 1972.

* About 23% of population in India's metros lives in slums. Dharavi in Mumbai, the largest slum in Asia, is home to 500,000 people in a small area of 170 hectares.

CURRENT ENVIRONMENTAL SCENARIO

The unprecedented population increase, and anthropogenic activities such as industrialization and urbanization of the twentieth century in the name of modernization have not only increased conventional solid and liquid waste pollutants to critical levels but also produced a wide range of previously unknown contaminants in the form of xenobiotics for which society is unprepared. The world population is expected to reach 8 billion by 2020. Population growth with global industrialization and successful green revolution has placed major pressure on our environment, like a boomerang, potentially threatening the sustainability. They have deleterious effects on living beings throughout the biosphere most notably in soils and sediments. Majority of the contaminants entering into the ecosystems are in the form of chemicals and they affect man, animal life, plant life and microbes and exert serious health and ecological problems. Chemicals and wastes include pesticides (agriculture), metals (industry), plasticizers and stabilizers (replacing traditional materials) and domestic (household) wastes. More than 60,000 chemicals are used in our day-to-day life in the form of fuels, industrial solvents, drugs, pesticides, fertilizers, food additives and consumer products. Besides these, we are adding every year as many as 500–1,000 new chemicals to the existing array of chemicals (Sharma, 2000). It has been identified that more than four million chemicals are either isolated from natural products or synthesized artificially.

Though rapid industrialization and faster growth rate are the requirements of a developing country, a holistic approach with environmental consideration is essential for sustainable development. Sustainable industrial development with environmentally sound and cleaner production is the challenge for the next century. "Environmental protection" that includes conservation of resources has three primary objectives.

1. Prevention of damage and discomfort.

2. Improvement in productivity and pleasure.

3. Maintenance of the ecosystem.

Microbes, the oldest inhabitants of the earth that are versatile and adaptive to the changing environment, will be the cost-effective components to combat the present problems. Microbes and their diverse metabolic enzymes are typically employed for safe removal of environmental contaminants, either through direct destruction or indirectly through a transformation of the contaminant to a safer intermediate. The limitations faced by physical and chemical treatment technologies will be overcome with the help of microbes. The United Nations Environment Programme (UNEP) defines the cleaner production concept as "the continuous application of an integrated preventive environmental strategy to processes, products and services to increase eco-efficiency and reduce risks to humans and environments (UNEP, 1996)".

> *Population growth with global industrialization and successful green revolution has placed major pressure on our environment, like a boomerang, potentially threatening the sustainability.*

Environmental Issues and the Public

People are either not aware of or not serious about environmental issues due to several reasons which include the following.

1. For millions of poor people, the problems of daily existence are more important than environmental deterioration.

2. The more affluent people are afraid that in the name of environment, their comforts may be lost.

3. False anti-environmental propaganda is given by vested interests like the large corporations.

4. At political level, parties are more interested in short-term gains and will not support unpopular measures to conserve the environment.

5. At this age of extreme specialization, very few can look at the larger picture of what is happening around the world.

6. There is paucity of reliable and clear information on the environment.

7. Indications like Gross Domestic Product (GDP) give us a misleading picture of what is desirable.

GLOBAL CONCERN ON ENVIRONMENTAL CRISIS

Rachel Carson's book *Silent Spring* published in 1962 may have set the first tone for an environmental movement in the twentieth century. The book describes how the use of pesticides, DDT in particular, affected people's health and also destroyed wildlife to such an extent that the spring arrived without the song of the birds. It appeared right at the beginning of the environmental movement and became a cult book (Rajagopalan, 2005). The solid scientific basis of *Silent Spring* led to the banning of DDT in the U.S. The United Nations Conference on Human Environment in 1972 at Stockholm was the first international initiative to discuss environmental problems. The report (1983) of World Commission on Environment and Development (WCED), *Our Common Future,* emphasized the need for an integration of economic and ecological systems. The commission supported the concept of *sustainable development* and defined it as "development that meets the needs of the present without compromising the ability of future generations to meet their own needs". The other major effort was the United Nations Conference on Environment and Development (UNCED) held in 1992 in Rio de Janeiro. It came up with the following documents.

- o "The Rio Declaration on Environment and Development" listing 27 principles of sustainable development.

- o "Agenda 21", a detailed action plan for sustainable development in the twenty-first century.

- o "The Convention on Biological Diversity."

> *The limitations faced by physical and chemical treatment technologies will be overcome with the help of microbes.*

The next conference popularly known as Rio+10, held in Johannesburg, South Africa, in 2002 recognized that the implementation of the Rio agreements had been poor. It marked a shift from agreements in principle to more modest but concrete plans of action.

Stockholm Convention on Persistent Organic Pollution (POP)

Stockholm Convention of December 2000, represented by 122 countries finalized a treaty to ban 12 chemicals, the so-called **"Dirty dozen"**. Among them, 8 are pesticides (aldrin, chlordane, DDT, dieldrin, endrin, heptachlor, minex and toxaphene), 2 are industrial chemicals (PCBs and hexachlorobenzene) and 2 by-products of combustion and industrial process (dioxins and furans). This treaty was the first-ever global agreement to abolish a class of chemicals. Most of the dirty dozen have already been banned in industrialized countries, but were still used in India and Latin America (Hill, 2004).

UTILIZATION OF WASTE BIOMASS RESOURCES

Waste is any material that is not needed by the owner, producer or processor. It has always been a part of the Earth's ecosystem. In fact, there is no real waste in nature. Waste from one process becomes an input for another. But exponential growth of human activities has created a situation where wastes need to be managed because they pose a problem to the environment and cannot be handled by nature. The composition, quantity, and method of disposal determine its adverse effects. Wastes have to be recycled, permanently isolated for storage, allowed to decompose and degrade into a harmless state, or treated to remove any toxicity they may have.

> *Waste from one process becomes an input for another.*

Current Situation on Waste Dispersal

Environmental policy for the dispersal of industrial waste is based on the command-and-control (C and C) strategy, by which the authorities concerned may impose fines and eventually close down non-complying enterprises. The solid and liquid wastes and gas emissions, characterized as hazardous, and sometimes toxic, generated by industrial activities are mainly disposed of in the sewage systems, rivers and streams, waterways, coastal lagoons and estuaries, open air dumps and so on. Large enterprises do not always move from the "end of the pipe" to the "front of the pipe", which would imply changes in technological processes sometimes requiring heavy investment. Enterprises are not yet moving sufficiently from environmentally dirty technologies to cleaner or less "dirty", ones. In recent years there is a paradigm shift from pollution control to pollution prevention (PP), an approach that can be adopted within all sectors. This can also be called waste minimization, waste avoidance, waste reduction, waste prevention, green productivity, source reduction, eco-efficiency, cleaner production, industrial ecology, green productivity, and so forth.

Five R Policies for Waste Minimization

Van Berkel (1995) suggested that the vital working method at the plant level for cleaner production includes the five R Policies, the main component of waste minimization strategy. It includes:

1. process modifications aimed at the **reduction** of waste
2. feedstock substitutions seeking the **replacement** of toxic raw materials for environment-friendly inputs
3. efficient use of water, energy or inputs by means of **reuse** and **recovery** practices
4. **recycling** and **recovery** of wastes in bioprocesses

Current bioremediation processes are mainly based on the main components of 5R policies that include reduction, replacement, reuse, recovery and recycling.

Strategies for Cleaner Environment

Environmental training and information programmes are needed, in which both the public and the private sectors should participate. The important methods useful in the management of contaminants in the environment include:

1. in-process treatment
2. end-of-pipe treatment
3. remediation of polluted sites

4.   modification of existing processes

5.   introduction of new processes and products

The first three options are targeting clean-up/removal of the pollutant and the remaining two are long-term strategies.

Choice of the Technology

Though many technologies have been available for clean-up purpose, only a few of them have been proved to be of routine application value. The choice of technology is influenced by the social, political, and geographical conditions. The technologies which satisfy most of the requirements and fulfil our goals, become our choice. The available technologies can be categorized as potential and time-proven trusted technologies. The available technologies are assessed based on certain criteria.

1.   whether it offers a temporary solution or permanent one

2.   whether it brings about more conversion or total elimination

3.   cost-effectiveness

4.   time taken for treatment

5.   whether the effluent then meets the regularity standards

> *The choice of technology is influenced by the social, political, and geographical conditions. The technologies which satisfy most of the requirements and fulfil our goals, become our choice.*

Bioremediation Techniques

Among the biological techniques, bioremediation has evolved as the most promising one because of its economical, safety and environmental features since organic contaminants become actually transformed and some of them are fully mineralized by this technique (Saval, 2003). The bioremediation process involving the use of microbes to detoxify and degrade environmental contaminants has received increasing attention as an effective biotechno-logical approach to clean up a polluted environment (Khan and

Anjaneyulu, 2005). Intrinsic bioremediation is the microbial removal of contaminants from the environment without human intervention and the process is very slow. Bioremediation is strongly limited by biodegradation and biotransformation.

> *Bioremediation is strongly limited by biodegradation and biotransformation.*

Biodegradation

Physico-chemical degradation processes differ from biodegradation and bioconversion. The physico-chemical processes involve chemical conversions such as photolysis, oxidation and reductions. However they do not involve the biocatalysts known as enzymes. The latter, i.e., metabolism, involves chemical conversions in a biological system carried out under the influence of enzymes, other than the body constituents. Efficient contaminant degradation involves various factors such as bioavailability, diffusion and transportation of the substrates into the cells, presence of the degradative bacteria at the site to be bioremediated and the ability of the organism to perform degradation under manipulated environmental conditions. Biodegradation involves complete breakdown or mineralization of molecules to carbon dioxide and water. For example, the filamentous marine fungus, *Cunninghamella elegans,* transforms PAH benz(α)anthracene to *trans*-dihydrodiols and related compounds biologically, which accumulate and are not mineralized (Cernigila *et al.,* 1985).

Biotransformation

Biotransformation refers to any transformation of the structure of a compound by living organisms or enzymes. The metabolic transformation of toxicants is known as detoxification because they are converted to less toxic products. The organisms also convert the toxicants into more toxic substances and the transformation of such a compound is activation. In living organisms, the polar or hydrophilic compounds are eliminated quickly. However non-polar or biopophilic compounds are gradually absorbed and accumulated as they are not easily eliminated. The lipophilic substances are altered into the hydrophilic forms biocatalytically so as to enable easy excretion. Biotransformation can convert an active chemical into an inactive one and vice-versa. Enzymes found in the microorganisms catalyse biotransformation reactions.

Implementation of Bioremediation Programmes

New factors such as globalization and competitiveness are forcing waste-generating industries to voluntarily seek certification programmes such as Environmental Management Systems regulated by international standards. Such programmes include management systems for environmental auditing and for environmental performance evaluation. The wide spectrum of standards force industries to seek new processes and products even if it is at the cost of new investment. This is a very good alternative for regulation of the industries by governments (rigid command-and-control legislative approach).

ENVIRONMENTAL CONTAMINANTS

EARTH DAY

The first Earth day was organized by Denis Hayes in US on April 22, 1970 which was founded by Gaylord Nelson. Twenty million Americans participated in the event and it marked the beginning of the environmental movement and a clear focus on reducing pollution. The Earth Day Network (EDN) promotes environmental awareness and it reaches over 12,000 organizations in 174 countries. More than half a billion people participate in the campaigns every year. Earth day has been an annual event for people around the world to celebrate the earth and our responsibility toward it. The official date of Earth day is 22 April.

ENVIRONMENTAL CONTAMINATION VERSUS POLLUTION

Among 87,000 commercial chemicals, the United States Environmental Protection Agency (USEPA) has identified 53 chemicals as Persistent, Bioaccumulative and Toxic (PBT). Of these, 42 are polychlorinated organic chemicals and the other eleven are metals. They are the major environmental pollutants that have led to serious surface and subsurface soil contamination. **Contamination** is the presence of elevated concentrations of substances in the environment, food, etc., which may not necessarily be harmful or a nuisance. **Pollution** can be defined as the introduction by humans, deliberately or inadvertently, of substances or energy (heat, radiation, noise) into the environment resulting in a deleterious effect. Pollution involves contamination, but contamination need not constitute pollution (Deswal and Deswal, 2004). Contaminants and pollutants are produced due to natural activities and anthropogenic activities such as large-scale manufacturing, processing and handling of chemicals as in petroleum-industry. **Xenobiotics** are the chemicals synthesized by humans that have no close counterparts. Polychlorinated biphenyls (PCBs), trichloroethylene (TCE), polyaromatic hydrocarbons (PAHs) and pesticides are considered as xenobiotics because of their recalcitrance to biodegradation and they have the capacity to get accumulated in living systems.

> *Pollution involves contamination, but contamination need not constitute pollution.*

NATURE OF CONTAMINANTS

The variety of materials and processes used in modern-day industrial activities cause different types of contaminants. The number of contaminants found to date are enormous and the types of mixtures are countless. Most of these chemical are produced and used to improve human health, standards of living and safety through advancements in manufacturing, agriculture, and medicine and strengthening national defence. Their unplanned intrusion into the environment can reverse the same standards of living that they are intended to foster.

Recalcitrant Compounds

The compounds which are not degraded in nature following their release into the environment even when conditions appear to be adequate for microbial growth are termed as recalcitrant. These compounds persist in all natural environments regardless of whether they are inherently biodegradable or not. Agro- and petrochemicals form a significant percentage of recalcitrant compounds and they disrupt the environment. Recalcitrant compounds have their half-lives which indicate the extent of persistence. It is estimated that 150×10^6 tonnes y^{-1} of recalcitrant organic chemicals are produced annually which find their way into the soil, surface water and ground waters as constituents of industrial effluent (Jogdand, 2004).

> ***Xenobiotics*** *are the chemicals synthesized by humans that have no close counterparts.*

Pollutants

Under the Clean Water Act (CWA), any material which can contaminate water is a pollutant and includes dredge soil, solid waste, incinerator residue, sewage, garbage, sewage sludge, munitions, chemical wastes, biological materials, radioactive materials, heat, wracked or discarded equipment, rock, sand, cellar dirt and industrial, municipal and agricultural waste discharged into water. Several types of common pollutants are categorized as shown in Table 2.1.

Table 2.1 Categories of common pollutants

Category	Examples
Organic chemicals	PCBs, oil, pesticides
Inorganic chemicals	Salts, nitrates, metals and their salts
Organometallic chemicals	Methyl mercury, tributlin
Acids	Sulphuric, nitric, hydrochloric
Physical	Eroded soil, trash
Radioactive	Radon, radium, uranium
Biological	Microbes, pollen

Nature of Contaminants in an Environment

A contaminant can exist in different forms (Figure 2.1). For instance in water, a contaminant can exist in three different forms which affect its availability to organisms (bioavailability).

1. dissolved
2. absorbed to a biotic compound and suspended in the water column or deposited on the bottom
3. incorporated (accumulated) in organisms

Water-soluble contaminants may persist and retain their physical and chemical characteristics. The persistent contaminants can accumulate in the environment to toxic levels. The persistence of a chemical contaminant may be expressed in terms of its half-life, which is defined as "the time required to reduce the initial concentration of the chemical by one-half".

Physico-chemical Properties of Contaminants

It is essential to analyse the physico-chemical properties of contaminants to evaluate its potential for bioremediation. The properties include density, adsorption coefficient for soils, solubility in water, solubility in various solvents, volatilization from soil, volatilization from water, etc. (Troy, 1994). The important physico-chemical properties needed for efficient bioremediation are listed below:

- **Electron acceptors** Biological reactions that yield energy are redox reactions. Thus adequate electron acceptors are important to control bioremediation. Electron acceptors establish the metabolism and it is essential for specific degradation reactions. From thermodynamics consideration, the electron acceptors are preferred in a definite order: oxygen, nitrate, CO_2, and organic chemicals (Fletcher, 1994).

- **pH** Most of the microbiological activities will take place in the pH range 6–8. Cellular functions, cell membrane transport and the equilibrium of microbial catalytic reaction are controlled by pH.

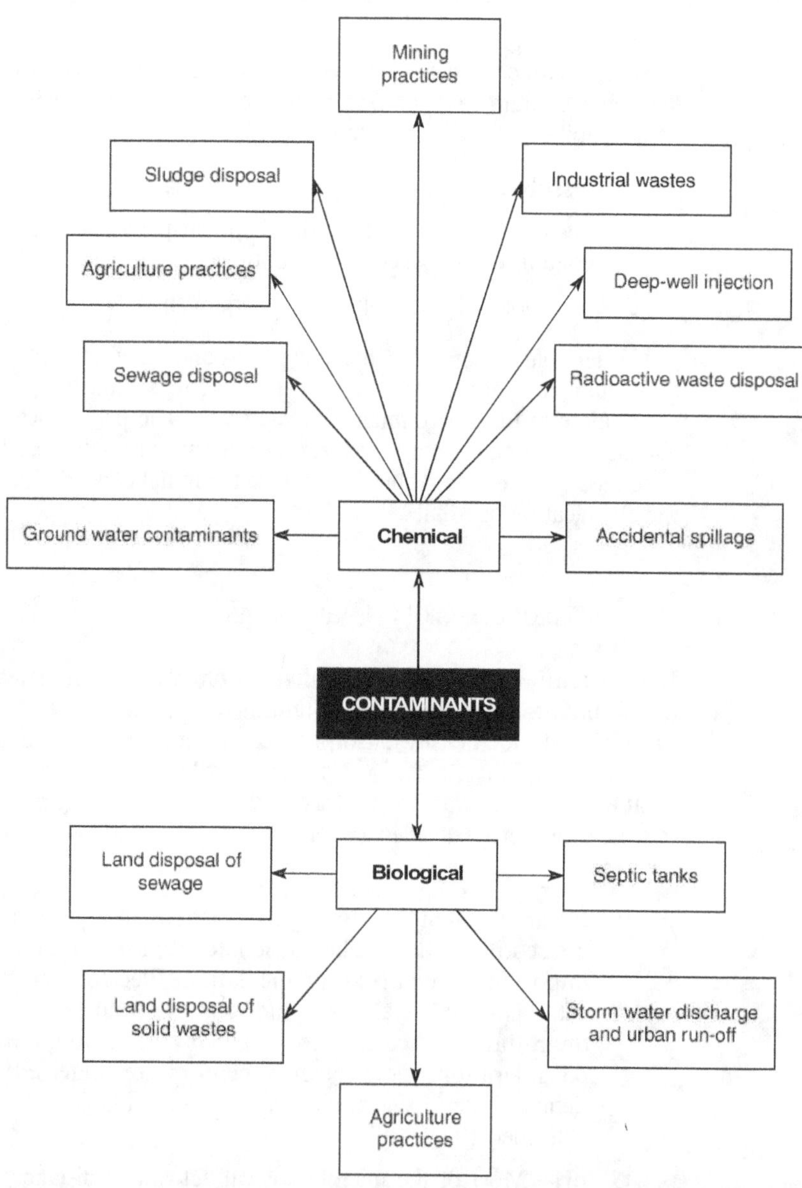

Figure 2.1 Sources of contaminants

o **Temperature**   The rate of degradation depends on the temperature. Microbial metabolism and enzymes can operate at specific ranges. The optimum temperature range for microbial volatilization is 10–40°C (Cookson, 1995).

o **Moisture**   Moisture content of soils affects the availability of contaminants, the transfer of gases, the effective toxicity level of contaminants, the growth and movement of microbes and their distribution. High water content influences oxygen transfer into the soil and limits growth efficiency. The optimal soil moisture content for biodegradation is 10–20% (King *et al.,* 1992).

> *The persistence of a chemical contaminant may be expressed in terms of its half-life, which is defined as "the time required to reduce the initial concentration of the chemical by one-half".*

GENERAL CLASSIFICATION OF CONTAMINANTS

Chemicals that are found in the soil system are a result of human activities. Contaminants are not found individually but in simple or complex mixtures. The mixtures may be associated with the release, storage or transport of many chemicals in surface or ground water, waste-treatment systems, soils or sediments. Concentrations of individual contaminants may vary significantly within the site, with some areas having extremely low concentration and others having concentration over a million fold higher (Trejo and Quintero, 2003). Contaminants can be classified (Figure 2.2) based on their

1. density

2. origin

3. chemical nature

4. degradation ability

5. health effects

6. environmental conditions

7. hazardous nature

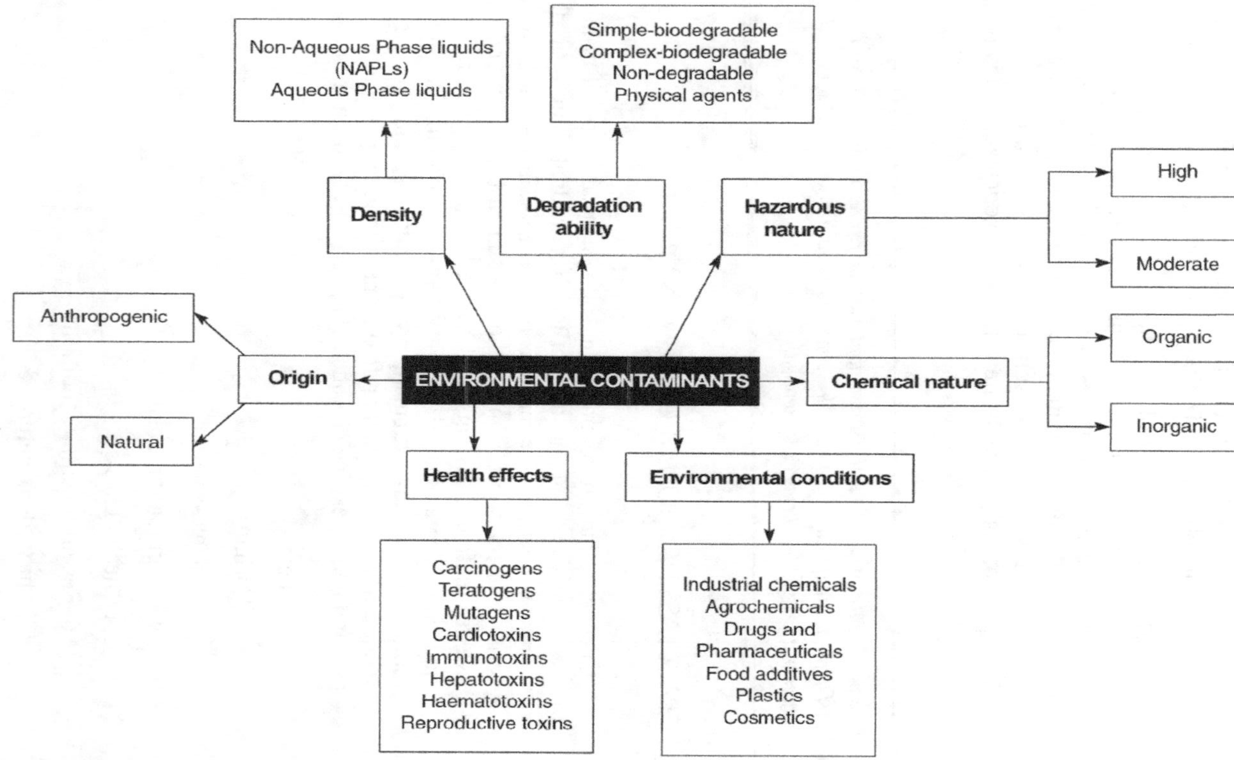

Figure 2.2   Classification of environmental contaminants

Contaminant Classification Based on Carbon Content

The contaminants are classified as organic and inorganic based on their carbon content that includes different resources (Figure 2.3).

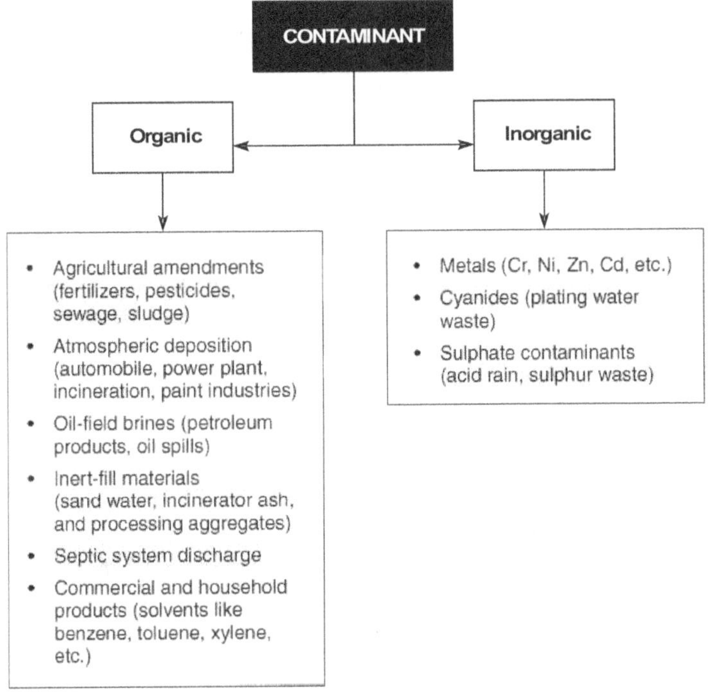

Figure 2.3　Classification of contaminants based on carbon content

Organic contaminants　An organic chemical contains the element carbon and it has carbon-to-carbon bonds that is, the molecule contains more than one carbon atom. If the chemical contains only carbon and hydrogen it is called a hydrocarbon. If carbon atom is bonded to a metal, the chemical is an organometallic. An organic compound can be synthetic, that is, synthesized from chemicals found in feed materials such as petroleum, coal, wood or cultures of molds or bacteria. An organic compound can be a petrochemical derived from crude oil or natural gas or synthesized using that oil or gas as a feed material. A biochemical is an organic chemical synthesized by microbes, plants or animals. **Aliphatic class** of organic compounds have the carbon backbone arranged in branched or straight chains (propane or octane). **Aromatic organic** compounds have a molecular structure with benzene (C_6H_6) as the basic unit (benzene, toluene, xylene and phenol). Organic contaminants can be classified as either aliphatic or aromatic compounds that contain deficient functional groups such as $-OH$, $-Cl$, $-NH_2^+$, $-NO_2$ and $-SO_3$. As electron donors, these contaminants undergo oxidation as a result of microbial metabolism (mineralization) or they form breakdown intermediates that may be utilized for microbial growth. Functional groups such as $-NH_2$, $-NO_2$ and $-SO_3$ may either be used as nutrients or cleaved from the carbon skeleton when the compound is reduced or oxidized. Oxidation process may be either aerobic or anaerobic and the fate of the contaminant is determined by the presence or absence of oxygen. Oxygen can act as a terminal acceptor of electron that is released during the oxidation of electron donors, or it can react directly with the organic molecule. As an electron acceptor, other oxidized inorganic compounds such as nitrate, metal ions, sulphate or carbon dioxide can replace oxygen, although the energy gains are smaller (Bouwer and Zehnder, 1993).

> *An organic chemical contains the element carbon and it has carbon-to-carbon bonds that is, the molecule contains more than one carbon atom.*

Inorganic chemicals   An inorganic chemical usually does not contain carbon although a few do, such as, sodium bicarbonate and sodium carbonate. Metals, such as copper, nickel, zinc and cadmium are often detected in contaminated soil in varying concentration. Higher concentrations result in the inhibition of metabolic activities of the microbes. Cyanide contamination results mainly from disposal of plating bath wastes and plating shop wastes. Cyanide ion is a non-specific enzyme inhibitor but exerts its powerful toxic effects by inhibiting the enzyme cytochrome oxidase, and this prevents the uptake of oxygen by living cells. Sulphate contaminant is mainly due to human activities with sulphur and its compounds.

Contaminant Classification Based on their Densities

Water-insoluble organic contaminants, i.e., non-aqueous phase liquids (NAPLs) that are produced by industrial activities are

> *An inorganic chemical usually does not contain carbon although a few do, such as, sodium bicarbonate and sodium carbonate.*

classified based on their density (Figure 2.4). Classification based on their density is important because when lighter-than-water non-aqueous phase liquids (LNAPLs) reach the water table, they tend to float on the water and spread rapidly, whereas denser-than-water non-aqueous phase liquids (DNAPLs) continue their downward path until the movement is arrested. DNAPLs are more hazardous because they pollute whole aquifers (Mackey and Cherry, 1989).

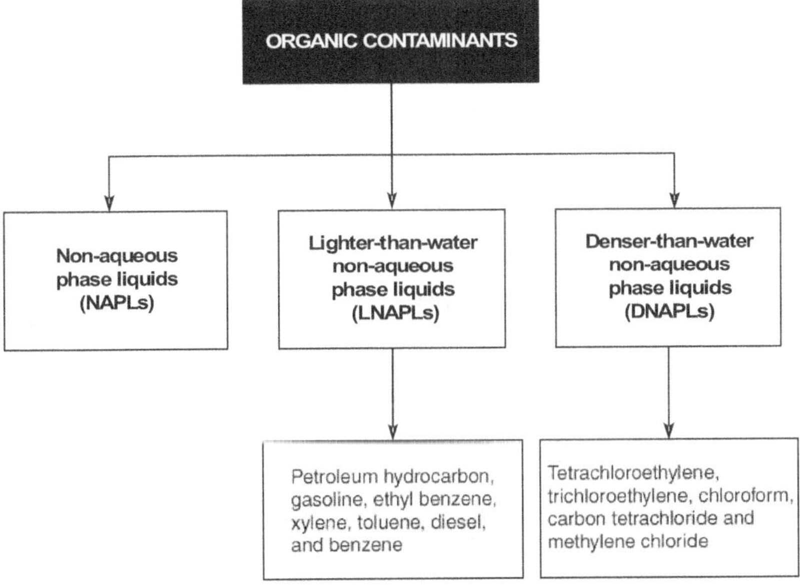

Figure 2.4 Classification of organic contaminants based on their density

EFFECTS OF CONTAMINANTS ON ENVIRONMENT

The physical and chemical properties of contaminants and other xenobiotics can have a profound effect on the biological activity of organisms found in an ecosystem. The effects of contaminants on the environment depends on several factors.

1. physical and chemical properties of chemicals and their transformation products

2. concentration of chemicals entering the ecosystem

3. duration and types of inputs (acute or chronic, intermittent spill or continuous discharge)

4. location of the ecosystem in relation to release site of the chemical

5. properties of the ecosystem that enable it to resist changes which could result from the presence of the chemical

> *Oxidation process may be either aerobic or anaerobic and the fate of the contaminant is determined by the presence or absence of oxygen.*

STRATEGIES FOR CONTAMINANT MANAGEMENT

Deterioration of the environment has become evident in the last few years as a result of inappropriate final dispersal of all sorts of waste materials. The variety of materials and processes used in industrial activities causes different types of contaminants in soil and they are not found individually, but in simple or complex mixtures. The numbers of chemicals found to date are enormous, and the types of mixtures are simply countless. Concentrations of individual contaminants may vary significantly within the site, with some areas having extremely low concentration and others having concentration over a million fold higher.

Hydrolysis, photolysis, oxidation, reaction with mineral surfaces and microbial degradation are some of the important environmental processes that may convert haloorganic and other organic compounds into non-toxic or more toxic compounds (Figure 2.5).

Hydrolysis   Water reacts with the functional groups attached to a hydrocarbon backbone such as chloro groups, esters, and other derivatives of carboxylic and phosphoric acids. Functional group hydrolysis makes the organic part of the molecule more soluble in water, hence more amenable to microbial degradation.

$$RX + HOH \longrightarrow ROH + HX$$

$$RCOX + HOH \longrightarrow RCOOH + HX$$

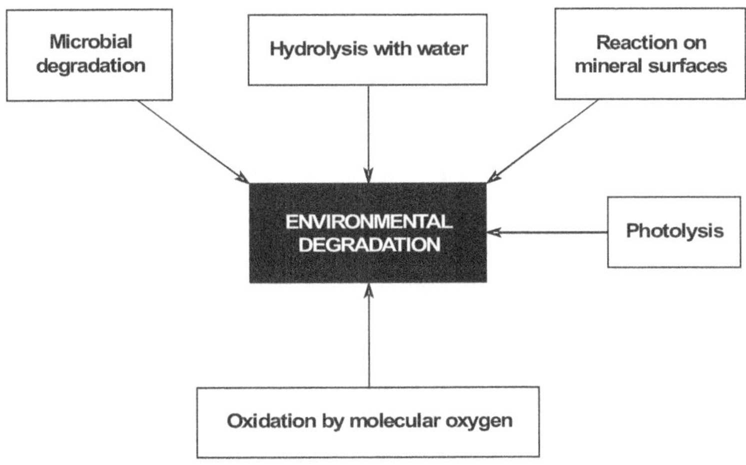

Figure 2.5   Environmental processes that may cause degradation

Reaction on mineral surfaces   Many organics bind to the surfaces of minerals in sediments. Some of these minerals catalyse the decomposition of substituted organics, e.g. sedimentary clay minerals catalyse the hydrolysis of chloro-organics.

Photolysis   The photochemical reactions of organics occur mainly with volatile organic compounds, especially the chemicals used in agriculture. Pesticides and herbicides are degraded by solar radiation. Volatile organics may also be decomposed by the radicals produced by the photolysis of other atmospheric gases.

Oxidation by molecular oxygen   Unsaturated compounds are more susceptible to oxidation by molecular oxygen. Sometimes the oxygen is activated by UV light to the singlet excited state in which it is more reactive than the triplet ground state oxygen.

Microbial degradation   Microbes are able to break down compounds that are not normally present in the environment into compounds that are a part of the natural biosphere. Such a transformation results in the conversion of a pollutant to a non-pollutant. Non-biological processes convert a non-biodegradable compound to one that is susceptible to degradation by microbes. Extremely refractory organic compounds due to their toxicity, complexity and inertness are resistant to microbial attack. This condition may manifest itself in the individual molecules or occur only above certain threshold concentrations. This may be due to several factors such as solubility, molecular size, amount of testing, branching and the number, nature and position of the substituents. Kaul and Szpyrkowicz (1998) listed the following general rules to determine the relative biodegradability or non-biodegradability of a compound.

o Materials that can pass through cell membranes are more readily available to the microbes and are degraded faster.

o Microbes prefer non-aromatic or cyclic aromatics to aromatics. The presence of the substituents in the benzene ring usually increases biodegradability.

o Soluble compounds are more easily degraded than insoluble compounds.

o A high degree of branching imparts greater resistance to degradation.

o A dispersed compound provides more surface area for attack, and therefore, degraded better.

o Compounds with unsaturated bonds are degraded more readily than saturated compounds.

o Materials such as alcohols, aldehydes, acids, esters, amines and amino acids are more preferred than the corresponding alkanes, alkenes, ketones, dicarboxylic acids, acids, nitrites, amines and chloroalkanes.

BIOAVAILABILITY OF CONTAMINANTS

The readiness with which a solute is in the reach of a potentially degradative microbe has been referred to as bioavailability. It depends on sorption (adsorption as well as absorption) and retention of the solute in the soil matrix. Bioavailability depends on the nature of the contaminant. For resistance, the organic fractions of soils and sediments have great affinity for hydrophobic compounds such as polycyclic aromatic hydrocarbons and other non-polar chemicals. Bioavailability is not a property of one chemical molecule alone but a property of one complex-molecule environment (e.g. soil moisture, salinity, aqueous phase, pH, etc.). Bioavailability is the main constraint to overcome (whether the contaminant is simple/more complex) before planning any bioremediation strategy. Contaminants persist in the environment due to the poor availability or the total lack of bioavailability to degrading microbes. Salkinoja–Saboren *et al.* (1990) summarized the following reasons for the persistence of organic chemicals in soil.

1. environmental incompatibility for microbial growth
2. insufficient nutrients for degraders
3. catabolic repression of degrading enzymes
4. unavailability of chemical compounds to microbes
5. biophysical factors

Bioavailability has been modelled in the traditional chemical reaction engineering way, i.e., through energy and mass balances coupled to reactions in which the soil–contaminant–microorganism is assumed to be the catalytic system. Chemical engineers have developed an interesting generic mathematical concept named bioavailability number (Bn), which resembles the inverse of the dimensionless Damkoller number (Bosma *et al.,* 1993). Calculation of Bn can give an idea of the local importance of mass transfer relative to the intrinsic activity of the microbes, e.g. Bn expresses control by microbial degradation at values greater than, and control by mass transfer at values less than, unity. In general, mathematical models fit well to experimental results when a simple contaminant is considered. However, models fail when complex mixtures of contaminants or soil are present.

Bioavailability Constraints

The major constraint in bioremediation technology is that every soil has different characteristics and as such no general rule exists for the microbes to readily adapt to new environment. The soils in various parts of the world have distinctive physical, chemical and biological characteristics that make them different from each other. The constraints that could influence bioavailability of contaminants for bioremediation strategy include:

a. Nature of microorganisms,
b. Composition and properties of contaminants and
c. Nature of the soil.

The decrease of the bioavailability in the course of time is often referred to as aging or weathering. It may result from

1. chemical oxidation reaction incorporating contaminants into natural organic matter

2. slow diffusion into very small pores and adsorption into organic matter

3. the formation of semi-rigid films around non-aqueous-phase-liquids (NAPL) with a high resistance towards NAPL water mass transfer

Soil Composition on Bioavailability

Soil composition is a constraint for the bioavailability of contaminants because it regulates

1. the maximum attainable capacity (partition coefficients) of pollutants in the liquid and solid phase

2. the kinetic and desorption rate of chemicals

3. the aggregation level, grain size and porosity of particles in which contaminants are trapped

The soil texture influences the water regime, gas exchange, local temperature, effective diffusion and micropore tortuosity.

Strategies to Increase Bioavailability

The general strategies that have been developed to increase bioavailability are the use of biosurfactants, addition of organic solvents and synthetic surfactants.

Bioavailability depends on the nature of the contaminant.

Surfactants   Surfactants have the capacity to reduce the surface tension of liquid media. Zhang and Miller (1992) postulated that if surface tension is reduced, it results in an increase of aqueous dispersion of non-ionic organic contaminants (NOCs), consequently solubility and rates of dissolution are also increased. Surfactants produced by microbes naturally are known as biosurfactants. Biosurfactants are exopolymers (polysaccharides, gums) mainly produced by bacteria under growth-limiting conditions (e.g. high carbon-to-nitrogen ratios and iron limitation) or when poorly soluble substrates such as n-alkenes are present. The use of biosurfactants is attractive because they are natural products and biodegradable. Among the most studied biosurfactants are rhamnolipid produced by *Pseudomonas aeruginosa.* Microbes naturally produce biosurfactants as a part of their strategy for growing on substrates where bioavailability is essentially reduced.

> *The use of biosurfactants is attractive because they are natural products and biodegradable.*

Organic solvents   Organic solvents change the polarity of the soil–water environment, influencing partition and consequently bioavailability of non-ionic pollutants. Appropriate organic solvents disperse NOCs up to molecular level, promoting a thin layer interface to which NOC degraders could adhere. Efronyson and Alexander (1991) proposed that solvents promote adherence of organisms to the interface in the neighbourhood of dispersed NOCs.

Synthetic surfactants   Synthetic surfactants are used to promote bioavailability of insoluble persistent contaminants such as PAHs and PCBs in soils and sediments. Synthetic surfactants increase both the apparent solubility and the maximum salts of dissolution.

> *Organic solvents change the polarity of the soil–water environment, influencing partition and consequently bioavailability of non-ionic pollutants.*

Rates of biodegradation of naphthalene and phenolphthalene were enhanced by the addition of surfactants in the dissolution-limited growth phase. Mineralization of the said compounds were unaffected by the addition of non-ionic surfactants even above their critical micelle concentration (CMC). Non-ionic surfactants at low concentration may promote the biodegradation of adsorbed aromatic compounds in polluted soils, even when surfactant-induced desorption was not appreciable.

BIOREMEDIATION

3

PSEUDOMONAS—POTENTIAL CANDIDATE FOR BIOREMEDIATION

Members of the genus Pseudomonas, a soil bacterium, are the most predominant microbes that degrade xenobiotics. Different strains of Pseudomonas that are capable of detoxifying more than 100 organic compounds, have been identified, e.g. hydrocarbons, phenols, biphenyls, PCBs, polycyclic aromatics and naphthalene. About 40–50 strains of microbes capable of degrading xenobiotics have been isolated. Sometimes for the degradation of a single compound, the synergetic action of a few microbes, i.e., a consortium or cocktail of microbes may be efficient. For example, the insecticide parathion is more efficiently degraded by the combined action of Pseudomonas aeruginosa and Pseudomonas stutzeri.

Satyanarayana (2005)

LAND FARMING IN SOIL BIOREMEDIATION

Land farming is a technique for the bioremediation of hydrocarbon-contaminated soils. The soil is excavated, mixed with microbes and nutrients and spread out on a liner, just below the polluted soil. The soil has to be ploughed regularly for good mixing and aeration. Addition of co-substrates and anaerobic pretreatment of polluted soils also increases the degradation process. The land farming technique has been successfully used for the bioremediation of soils polluted with chloroethane, benzene, toluene and xylene (BTX-Aromatics).

Satyanarayana (2005)

GENERAL PERSPECTIVES

In the course of the last two decades a wide variety of technologies had been developed for clean-up operations of contaminated sites. They can be classified as physical, chemical, thermal and biological. Among the biological techniques, bioremediation has evolved as the most promising one because of its economical, safety and environmental features, since organic contaminants become actually transformed, and some of them are fully mineralized (Susana Saval, 2003). Bioremediation technologies offer a cost-effective, permanent solution to clean up soils contaminated with xenobiotic compounds. According to Alper (1983), bioremediation is at least six times cheaper than incineration and three times cheaper than confinement. It is a new and exciting field and its multidisciplinary nature is a challenge for those interested in the remediation of contaminated sites. Bioremediation can be performed off-site when contamination is superficial, but it will have to be *in situ* when the contaminants have reached the saturated zone. The general components and characteristics of bioremediation constitute three important aspects.

- o microbial systems
- o type of contaminant
- o geological and chemical conditions at the contaminated site

The areas of research include the development of methods for increasing the bioavailability of persistent materials.

1. development of better qualitative and quantitative methods for sampling and analysis of the contamination.

2. improved and standardized techniques for assessment of biodegradation methods and microbial systems.

3. standardized methods to determine levels of toxicity and monitoring toxic compounds during and at the end of the treatment.

Definitions

Bioremediation is defined as the use of biological treatment systems to destroy, or reduce the concentration of hazardous wastes from contaminated sites (Trejo and Quintero, 2003).

The American Academy of Microbiology defined Bioremediation as the use of living organisms to reduce or eliminate environmental hazards resulting from accumulation of toxic chemicals and other hazardous wastes.

Bioremediation can also be defined as the complete removal of pollutants and their toxicity through the metabolic reaction mediated by microorganisms.

> *Bioremediation is at least six times cheaper than incineration and three times cheaper than confinement.*

It is a technology that utilizes the biological activity to reduce the concentration or toxicity of pollutant. It commonly uses processes by which microorganisms transform or degrade chemicals in the environment.

Bioremediation is the manipulation of living systems to bring about desired chemical and physical changes in a confined and regulated environment (Cacciatore and Mc Neil, 1995).

Bioremediation is often used to describe a variety of diverse microbial processes that occur in natural ecosystems, such as mineralization, detoxification, co-metabolism and activation (Rochelle *et al.*, 1989).

Bioremediation Approaches

Bioremediation is an organic approach to the reclamation of waste materials at the site. It is simply a new application of a very old technology once primarily used in waste-water treatment. Now this technology is routinely applied to soils, sludges, ground water and surface water contaminated with chemicals such as crude oil, petroleum, hydrocarbons, fuels, industrial solvents, etc. Detoxification and mineralization of the pollutants to biomass, CO_2 and water make it an attractive, environment-friendly, safe and cost-effective alternative technology to conventional methods. It is a strategy or a process that uses microorganisms, plants or microbial or plant enzymes to detoxify the contaminants in soils and other environments. The concept includes biodegradation, mineralization and co-metabolism. Three important basic principles have to be considered before selecting the most appropriate strategy to treat specific contaminants: the amenability of the contaminant to biological transformation to less toxic products (biochemistry), the accessibility of the contaminant to microbes (bioavailability) and the opportunity for optimization of biological activity (bioactivity).

Advantages and Disadvantages of Bioremediation

Bioremediation is a versatile process because it can be adapted to suit the specific needs of each site. Bioremediation is still considered an innovative technology that has been used in limited number of cases with several advantages in spite of certain disadvantages.

Advantages

- Can be done on site.
- Minimum site disruption is caused.
- Eliminates transportation costs and liabilities.
- Eliminates long-term liabilities.
- Uses biological system, often less expensive.
- Can be coupled with other treatment techniques.

Disadvantages

- Some chemical compounds are not biodegradable.
- Extensive monitoring is required.
- Each site has specific requirements.
- Potential production of toxic unknown sub-products is possible.
- Strong scientific support is needed.

Bioremediation is a versatile process because it can be adapted to suit the specific needs of each site.

Types of Bioremediation

Several biological systems and techniques have been developed in the last few years to clean the environment. Bioremediation explores the genetic diversity and metabolic versatility of microbes for the transformation of contaminants into less harmful end products, which are then integrated into natural bio-geochemical cycles (Liu and Suflita, 1993). The diversity of bioremediation technologies is an indication of variation in the hazardous environmental contaminants. The selection of bioremediation system depends on the constituent of the contaminants and it involves a multidisciplinary approach requiring knowledge from individuals with expertise in chemistry, microbiology, geology, environmental engineering, chemical engineering, soil science, etc. The bioremediation treatment technologies can be categorized as follows.

Bioaugmentation   Addition of bacterial cultures to a contaminated medium: frequently used in bioreactors and *ex situ* system. It is used when the proportion of degradative microbial flora to contaminant needs to be increased.

> *Detoxification and mineralization of the pollutants to biomass, CO_2 and water make it an attractive, environment-friendly, safe and cost-effective alternative technology to conventional methods.*

Biomineralization (Biocrystallization)   The process whereby microbially generated ligands or microbially mediated changes in the cellular microbial environment cause precipitation of heavy metals as biomass-bound crystalline deposits.

Biostimulation   Stimulation of indigenous microbial population in soils and / or ground water (*in situ* or *ex situ*). It can be applied when the addition of nutrients is necessary.

Bioattenuation   A method of monitoring the natural program of degradation to ensure that contaminant concentration decreases with time at relevant sampling point.

Bioventing   Treating contaminated soils by drawing oxygen through the soil to stimulate microbial growth and activity. It is needed when it becomes necessary to supply oxygen from air.

Biofilters   Use of microbial stripping columns to treat air emissions.

Bioreactors A container/reactor used to treat liquid or slurries.

Composting Aerobic, thermophilic treatment process in which contaminated material is mixed with a bulking agent.

Land farming Solid-phase treatment system for contaminated soils; may be done *ex situ* or in a constructed soil treatment cell.

The diversity of bioremediation technologies is an indication of variation in the hazardous environmental contaminants.

Factors Affecting Bioremediation

The key element in any bioremediation process is the aerobic or anaerobic heterotroph whose activity is affected by a number of physico-chemical environmental parameters namely, energy sources (electron donor), electron acceptors, pH, nutrients, temperature, moisture and inhibitory substances or metabolites. Boopathy (2000) has listed some important factors that may affect bioremediation (Figure 3.1).

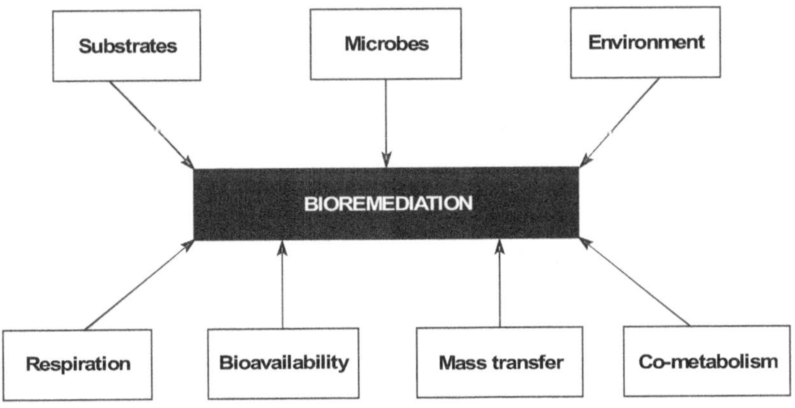

Figure 3.1 Factors affecting the process of bioremediation

The success of bioremediation techniques is directly related to the metabolic capability of involved microorganisms and can be affected by the surrounding environment. In addition to the above factors, non-ionic organic contaminants (NOC) have low aqueous solubilities, low dissolution rates and they are, as soil contaminants, strongly bound to or adsorbed onto solids. The biodegradation of such compounds in the natural environment may be restricted because:

1. the native population of microorganisms is absent or extremely poor and bioreaction limits the process.

2. the compounds to be degraded are not available to microorganisms and mass transfer becomes limiting.

3. compounds or their metabolic intermediates are toxic to the native microorganisms and biodegradation is inhibited.

> *The selection of bioremediation system depends on the constituent of the contaminants and it involves a multidisciplinary approach requiring knowledge from individuals with expertise in chemistry, microbiology, geology, environmental engineering, chemical engineering, soil science, etc.*

Rogers *et al.* (1983) pointed out the important characteristics essential to outline an ad hoc strategy for bioremediation.

1. chemical nature of the contaminant

2. whether contamination is superficial/subsoil

3. records to show biodegradability of contaminants

4. depth and extension of contaminant plume

5. depth of water table

6. nature of the permeability of the geological material

7. suitable microbes that can degrade contaminants

8. nature of environment for microbial action

9. possibilities to "build" a bioreactor at the site

Bioremediation Mechanisms

The chief ways by which remediation may be accomplished include biosorption, bioaccumulation, reduction, solubilization, precipitation and methylation (Alexander, 1999). Some of the technologies are fully developed and are being practised.

Biosorption   Biosorption is a property of certain types of inactive, dead, microbial biomass to bind concentrate contaminants from even very dilute aqueous solutions, e.g. heavy metals. Biomass that exhibits this property acts as a chemical substance (ion exchanger) of biological origin. The cell wall structure of certain algae, fungi and bacteria is responsible for this phenomenon, which is a passive process that requires no energy. Biosorption often results from the formation of metal-organic complexes with constituents of microbial cell walls, capsules, or extracellular polymers synthesized and excreted by the organisms. It may result from the positively charged metallic cation being retained electrostatically by negatively charged functional groups in the walls, capsules or polymers. The mechanism of sorption and the cell constituents involved in sorption are usually unknown, although considerable research has been developed to trace the possible mechanisms. It involves mechanisms like ion exchange, chelation and complexation. Inorganic precipitation may occur by hydrolysis. Inorganic deposition occurs via adsorption by physical forces and ion entrapments in inter- and intra-fibrillar capillaries and spaces of the structural polysaccharide network as a result of diffusion through cell walls and membranes (Vasudevan *et al.,* 2001). Several active groups of cell constituents like acetamido group of chitin, structural polysaccharide of fungi, amine (amino and peptidoglycosides), sulphahydral and carboxyl groups in protein, phosphodiester (teichoic acid), and phosphate, and hydroxyl in polysaccharides participate in biosorption.

> *The success of bioremediation techniques is directly related to the metabolic capability of involved microorganisms and can be affected by the surrounding environment.*

Bioaccumulation   The active mode of accumulation of contaminants by living cells is usually referred to as bioaccumulation. This process is metabolically dependent and can be significantly affected by the contaminants especially the metals. The use of living systems certainly has application in a selected recovery system. However, the toxicity and complexity of many wastes may limit the use of living systems. The mechanisms by which bacteria actively accumulate include precipitation, intracellular accumulation and oxidation or reduction (Ahmad *et al.,* 2004). The concentration of a foreign substance within a biological system is dependent upon the rate of uptake, the duration of exposure and the rate at which it is being eliminated or reached upon by the system. The extent up to which a chemical shall be absorbed and bioaccumulated by an organism depends on its solubility in fats or its lipophilicity. A higher degree of correlation has been found between octanol water pollution co-efficient (log P_{ow}) and the extent of bioaccumulation (Neely *et al.,* 1986). Pollutants soluble in lipoid material are capable of forming complexes with macromolecules within the cell may be stored for long duration of time. Complex formation and dissolution in lipids and fats, which are storage products of cellular metabolism, these substances stay away from enzymatic reactions, which act on them (Kenaga and Goring, 1980).

> *The chief ways by which remediation may be accomplished include biosorption, bioaccumulation, reduction, solubilization, precipitation and methylation.*

Precipitation   Contaminants react with a product of microbial metabolism to yield water in soluble derivative. The removal of such precipitate constitutes remediation. Sulphides and phosphates are the common precipitates formed in microbes due to the production of H_2S from sulphates and inorganic phosphates from organic phosphate compounds. The H_2S/phosphate forms insoluble derivatives with a number of metallic ions. Microbial sulphate reduction for remediation is important in the use of constructed wetlands to remove metals from acid and coalmine drainage and other water polluted with toxic ions and sulphate.

Reductive halogenation   This is potentially important in the detoxification of halogenated organic contaminants. The halogen atom (such as chlorine) of the contaminant molecule is replaced by a hydrogen atom due to catalytic reaction of microbes. Two electrons are added, and the contaminant becomes less toxic and susceptible to further microbial decay than the parent compounds. In most cases this reaction does not generate energy but is an incidental reaction that may benefit the cell by eliminating the toxicity of the compound.

Reduction   Microbes can bring about the reduction of a wide array of inorganic anions and cations. Nitrate, sulphate and carbonate are the non-metallic anions that are reduced microbiologically. The reductions convert a higher oxidation state of the element to a lower one, e.g. Hg(II) to Hg(0), Fe(III) to Fe(II), Se(VI) to Se(0), Mn(IV) to Mn(II), V(VI) to V(IV) and As(V) to As(III). The reduction changes the toxicity, water solubility and mobility of the element. In other words, one oxidation state is toxic but the reduced form has less or no toxicity at environmental concentrations. An increase in water solubility and mobility can be exploited to bioremediate the insoluble form of an element in soil because the product of reduction would move out the solids into the water. A decrease in solubility of an element as a result of reduction can be used to remove it from surface to ground water. In some cases, the product of reduction is volatile, as in the reduction of nitrate to N_2 or Hg(II) to Hg(0), and the loss of the element from soil or water is (as with nitrate) or could be (as with Hg) the basis for bioremediation (Alexander, 1999). The reduction reaction of microbes are catalysed enzymatically, but some result from the excretion by the organism of a reducing compound that abiotically converts the ion or metabolic compound from a higher to a lower oxidation state. The enzyme-catalysed reactions serve as terminal electron acceptors permitting growth, much as O_2 functions for aerobes, simple organic molecules for fermentative microbes, sulphate for *Desulfovibrio*, nitrate for denitrifying bacteria and CO_2 for methanogenic bacteria.

Solubilization/Oxidation The oxidation reaction that has practical use is associated with the oxidation of sulphides, Fe(II) and elemental sulphur. Chemoautotrophic bacteria of the genus *Thiobacillus ferroxidans* use sulphides and Fe as its energy source present in the ores and produce H_2SO_4 and Fe(III). The acidification process can be used in a beneficial way under a controlled or managed process known as bioleaching (Ehrlich and Brierley, 1990).

Limitations of Bioremediation

Bioremediation is an alternative method to incineration, catalytic destruction, adsorption, physical removal and the destruction of contaminants. Integration and proper utilization of natural or modified microbial capabilities with appropriate engineering designs provide suitable growth in the contaminated environments. However, a gap exists between advances in laboratory research and commercial field application (Binulal *et al.*, 2003). The main constraints are;

1. lack of sufficient knowledge base to accurately predict pollution degradation rates and fates

2. designated sites as centres for bioremediation research and technology demonstration

Though several microbes have been designed for bioremediation, the physiological potential of microbial population to remediate environment of relevant size, heterogeneity, and variability has not been adequately tested.

MICROBES FOR BIOREMEDIATION

Successful bioremediation requires not only the knowledge of the microorganisms that can degrade a particular compound, but also an understanding of the pathways involved in the degradation both at physiological and molecular levels. (Ghosh *et al.,* 2005). Microbial diversity includes the diversity of bacteria, protozoans, fungi, and unicellular algae and constitutes the most extraordinary reservoir of life in the biosphere. The main component of diversity constitutes two elements: richness (highest diversity occurs in communities with many different species present) and evenness (equal abundance). The richness and evenness of bacterial communities reflect selective pressures that shape diversity within communities. It has been estimated that one gram of soil might contain 1,000–10,000 species of unknown prokaryotes and there is likely to be further diversity within each species. The cultivability of bacteria from natural habitats ranged from 0.001% in sea water to 0.3% in soil (Amman *et al.,* 1995).

Essential Characteristics of Microbes for Bioremediation

Genetic, biochemical and ecological abilities of microbes play a significant role in biodegradation. Faster kinetic rate is economical and biomass with slow specific growth rate responds more favourably to shock loadings. Appropriate reactor designs can control microbial system under various operating conditions to give effluents of acceptable quality. Microbial degradation depends on several conditions which can be summarized as follows.

1. Presence of microbes with the capacity to degrade the target compounds.

2. Substrate being accessible to the organism and capable of being used as energy and carbon source.

3. Presence of inducer to cause synthesis of specific enzymes for the target compounds.

4. Presence of an appropriate electron acceptor–donor system.

5. Ideal moisture and pH for microbial growth.

6. Nutrients for microbial growth and enzyme production.

7. Optimum temperature to support microbial activity.

8. Absence of toxic substance.

9. Microbes to degrade metabolic products.

10. Microbes to prevent transit of toxic intermediate products.

11. Ideal conditions to minimize competitive organisms for those conducting desired reactions.

Microbial Adaptation for Adverse Conditions

Microbes adapt to different micro-environmental conditions like pH, temperature and pressure even at extreme variations because of their wide biochemical versatility. Microbes are capable of developing tolerance to contaminants that induce a toxic effect on their vital functions. Microbes have the genetic information capable of producing enzymes involved in transformation of contaminants that are capable of tolerating extreme conditions. The organic contaminants can be transformed into different molecular entities that are retained by the cell without degradation.

Successful bioremediation requires not only the knowledge of the microorganisms that can degrade a particular compound, but also an understanding of the pathways involved in the degradation both at physiological and molecular levels.

Microbes Involved in Bioremediation

In the process of detoxifying the contaminants, the appropriate microorganisms flourish and carry out the metabolic activity using the contaminants as nutrients and as energy source or degrade them by co-metabolism. Substrates such as carbohydrates, proteins, fatty acids, methanol, etc., present in the wastes can act as electron donors in the oxidation–reduction reactions to produce energy and provide carbon for cell synthesis. Chemoheterotrophic microorganisms utilizing these substrates require electron acceptors such as oxygen, nitrate (NO_3^-), sulphate (SO_4^{2-}), or carbon dioxide (CO_2) for the oxidation process.

Specific chemoheterotrophic microorganisms utilize the oxygen (as electron acceptor) and organic substance (as electron donors) present in the waste that are under aerobic condition. Similarly under anoxic conditions the electron acceptor is nitrate (NO_3^-), and under aerobic condition it is either sulphate or carbon dioxide, which can oxidize organic substances (electron donor) present in the waste. In addition more energy is produced from oxidation–reduction reactions that occur under aerobic conditions. Therefore aerobic microorganisms have more energy for cell synthesis compared to anaerobic microbes. Then the production of microbial mass (sludge) is higher in aerobic and anoxic treatment processes than the anaerobic processes. The microbes can be classified based on the type of energy and carbon sources they use in cell synthesis and maintenance (Table 3.1).

Table 3.1 Microbial classification based on the type of energy and carbon source

Energy Source	Carbon source	
	Carbon dioxide	Organic substances
Light	Photoautotrophs (Higher plants, algae, photosynthetic bacteria)	Photoheterotrophs (Purple and green bacteria)
Chemical	Chemoautotrophs Energy derived from reduced organic compounds such as NH_3, NO_2^-, H_2S, reduced forms of sulphur (H_2S, S, $S_2O_3^{2-}$) or iron	Chemoheterotrophs Carbon and energy usually derived from the metabolism of a single organic carbon

Metabolic Process Involved in Bioremediation

Microbes that live virtually everywhere are the vital components for bioremediation. Microbial enzymes act as catalysts in degradative (catabolic) reactions that provide energy and material for synthesis of additional microbial cells. In general, the biochemical process can be divided into two groups: fermentation and respiration. These can be distinguished by the nature of the redox reaction on the basis of the terminal electron acceptor (CETS, 1995). Two types of metabolism exist, depending on the type of electron acceptor (Figure 3.2). If it has an organic origin, fermentation occurs whereas for inorganic compounds, the process will be respiration. In turn, there are two kinds of respiration: aerobic, when the molecular oxygen becomes the electron acceptor: and anaerobic, when oxidized inorganic compounds such as nitrates, sulphates or carbon dioxide are used.

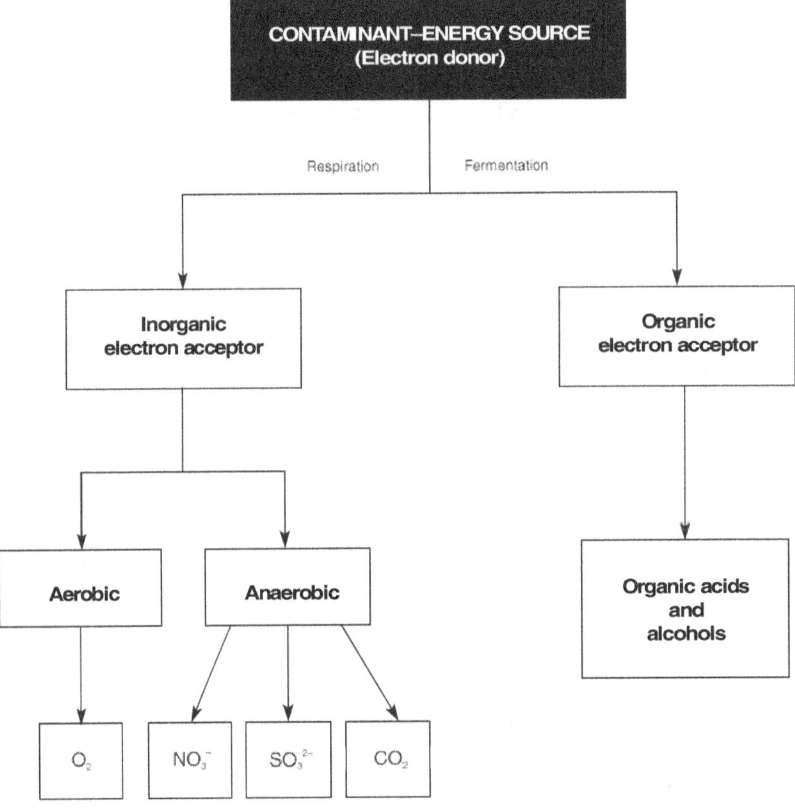

Figure 3.2 Metabolic routes based on electron acceptor

Fermentation   Microbes use the same organic compound as electron donor and acceptor. This process does not result in the complete oxidation of the substrate to carbon dioxide. Fermentation metabolism is characterized by production of a mixture of end products with different levels of oxidation, such as organic acids, alcohols, hydrogen and carbon dioxide.

Respiration   Respiration is of two types.

1. Aerobic, when the molecular oxygen becomes the electron acceptor. Microbes use oxygen to oxidize part of the carbon in the contaminant to carbon dioxide and the rest of the carbon is used to produce new cell mass.

2. Anaerobic, when the respiration occurs in the absence of molecular oxygen. Nitrate (NO_3^-), sulphate (SO_4^{2-}), iron (Fe^{3+}), manganese (Mn^{2+}), and even carbon dioxide serve as electron acceptors to degrade the contaminants. The by-products include nitrogen gas, hydrogen sulphide, reduced forms of metals, and methane, depending on the electron acceptor.

> *Microbes have the genetic information capable of producing enzymes involved in transformation of contaminants that are capable of tolerating extreme conditions.*

Co-metabolism (Synergistic transformation)   Microbes can transform contaminants even though the conversion reaction yields no benefit to the cell (Table 3.2). The non-beneficial bioremediation is termed co-metabolism/secondary utilization. It can be defined as degradation of a compound only in the presence of other organic material that serves as the primary energy source, e.g. polynuclear aromatic hydrocarbons, halogenated aliphatic and aromatic hydrocarbons and pesticides. A co-metabolite does not support the growth of the organism concerned and the products of transformation are accumulated stoichiometrically. The transformation does not require energy and it is a very slow process. Co-metabolism occurs, since the enzymes produced by organisms for metabolic activities of their major carbon source are not substrate-specific and can act on other compounds.

Table 3.2 Microbes involved in co-metabolism

Organism	Substrate	Co-metabolite
Methylococcus	Methane	Ethane
Nocardia	Hexadecane	Toluene
Achromobacter	Hexadecane	Ethyl benzene
	Benzoic acid	3-Chlorobenzoate
Cornyebacterium	Hexadecane	Naphthalene
	Glucose	Anthracene

Bacteria Versus Fungi for Bioremediation

Generally bacteria are easier to culture, and grow more quickly than fungi and are more amenable to straightforward molecular manipulation techniques. They are able to metabolize chlorinated organics and other organic contaminants better and they mineralize these chemicals and use them as carbon and/or energy source.

Fungi Polysaccharides in association with lipids and proteins, represent the main constituent of fungal cell wall. In filamentous fungi, the outer cell wall layer mainly contain neutral polysaccharides (glucans and mannoses), while the inner layers contain more glucosamines (chitin and chitosan) in a microfibrillar structure. Ligands within these matrices include carboxylate, amine, phosphor, hydroxyl, sulphahydral and other functional groups.

Bacteria The anionic nature of bacterial surface enables them to bind metal cations through electrostatic interactions. Cell wall thickness and anionic character, which is due to peptidoglycan, teichoic acid and teichuronic acid of gram-positive bacteria, have high capacity for metal binding.

Algae Special proteins and polysaccharides are present in the cell wall, that has potential for biosorption of contaminants. Polysaccharides of cell wall could provide amino and carboxyl groups and the nitrogen- and oxygen-based moieties could also form co-ordinated bond with the contaminants. Algin is present as a mixed salt of sodium, potassium, calcium and magnesium and is characterized as a high molecular weight polymer, that may be involved in the removal of contaminant.

> *The mechanisms by which bacteria actively accumulate include precipitation, intracellular accumulation and oxidation or reduction.*

Microbial Interaction for Bioremediation Optimization

Microbes exhibit positive (commensalism, synergism and mutualism) and negative (competition, amensalism, parasitism and predation) interactions that result in growth, survival and metabolism, that are essential for bioremediation.

Commensalism   Many contaminants are co-metabolized, a process in which microbes transform the contaminants without being able to use the energy derived from the metabolism to support growth. Co-metabolizing microbes may interact with other community members commensalistically, since co-metabolites serve as substrates for other microbes. Co-metabolism is the predominant mechanism for the transformation of many substrates, e.g. polychlorinated biphenyls (PCBs). Analogue enrichment and cross-accumulation are the bioremedial approaches developed after understanding co-metabolism. In analogue enrichment, ecosystem may be enriched for co-metabolizing microbes by the addition of biodegradable compounds. In cross-accumulation, microbes acclimated to the biodegradation of a particular substrate can be acclimated to another contaminant that is structurally related. These approaches are advantageous since they can reduce the length of time between the first exposure to the contaminant and the onset of biodegradation.

Synergism   Microbes interacting syntrophically often supply the nutritional needs of each other. Syntrophism is common in anaerobic microbial communities, possibly due to the small enthalpy changes associated with most anaerobic bioconversions. The excretion of metabolic end products with high free-energy content creates a niche for other organisms to exploit and this can result in community succession. Microbial interaction can have a thermodynamic basis, as in anaerobic microbial consortia that transfer hydrogen. Interspecific hydrogen transfer forms the basis for syntrophic interactions, e.g. metabolism of 3-chlorobenzoate.

Competition   Degradative microbes face competition with other organisms for nutritional and physical resources. Due to this competition, the expected bioconversion does not occur, even though the relevant degradative microbe is found in the environment. The dominance of one species over another can be due to its faster growth rate under the imposed selection pressure or its greater affinity for a substance, nutrient or physical resource. An understanding of the community structure is important for predicting the fate of contaminants in the environment and for remedial attempts.

BIOREMEDIATION TECHNIQUES

Bioremediation techniques are divided into three categories: *in situ, ex situ,* and *ex situ* slurry (Figure 3.3). With *in situ* techniques, the soil and associated ground water is treated in place without excavation, while it is excavated prior to treatment with *ex situ* applications. *Ex situ* slurry techniques involve the creation and maintenance of soil–water slurry as bioremediation medium. The slurry can be maintained either in a bioreactor or in a pond lagoon (Blackburn and Hafker, 1993).

> *Microbes can bring about the reduction of a wide array of inorganic anions and cations.*

In situ Remediation Techniques

In situ remediation includes pump-and-treat, percolation and bioventing/air spraying. Pump-and-treat system involves removal, treatment and return of associated water from a contaminated zone. The water is supplied with nutrients and saturated with oxygen. Percolation consists of applying water containing nutrients and microbial inoculum to the contaminated site. Bioventing is the process of supplying air to an unsaturated soil zone through the installation of wells connected to associated pumps. Air spraying involves the injection of air into the saturated zone of a contaminated soil.

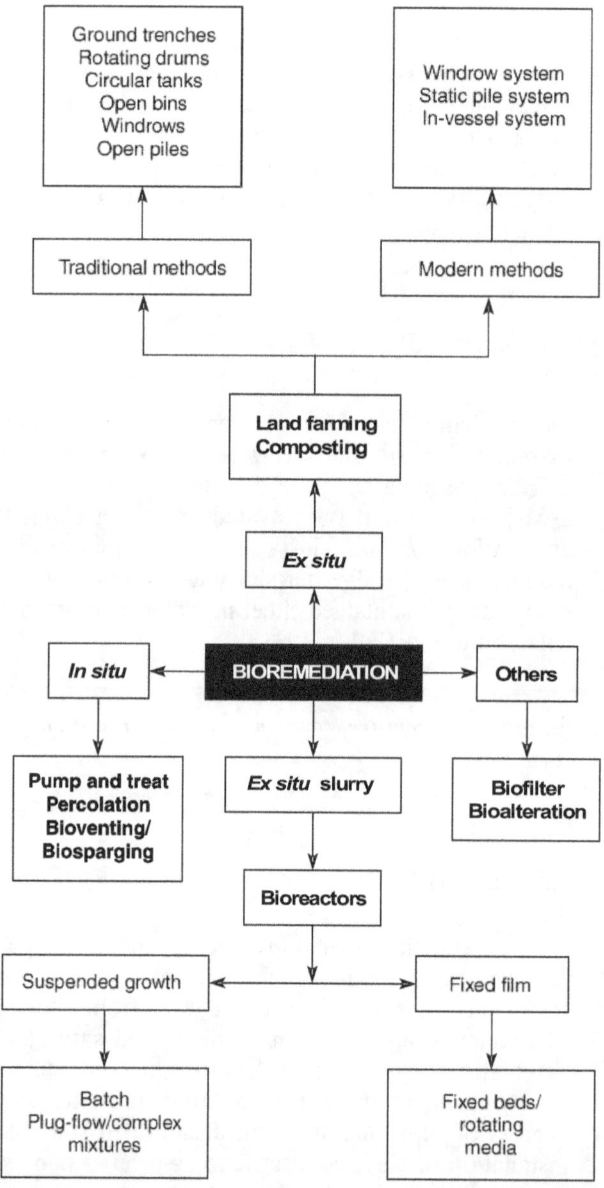

Figure 3.3 Bioremediation strategies to treat contaminants

> *Bioremediation techniques are divided into three categories: in situ, ex situ and ex situ slurry.*

Ex situ Techniques

Ex situ solid-phase technique consists of soil-treatment units, compost piles and engineered biopiles. Soil treatment consists of soil contained and tilled with application of water, nutrients and microbes to the soil. Compost piles consist of soil supplemented with composting materials to improve its physical handling properties and its air- and water-holding capacities. Biopiles are piles of contaminated soil that contain piping to provide air and water. Addition of water, nutrients and microbes are the main aspects of *ex situ* remediation.

Characterization of Essential Factors for Bioremediation

All the contaminated sites are not suitable for treatment with bioremediation techniques. It is essential to know information about three closely related aspects (Figure 3.4): the chemical nature of contamination, the geohydrochemical properties, and the biodegradation potential for the site (Heitzer and Sayler, 1993). Efficacy, reliability and predictability of the contaminated sites are essential in advance to assess the suitability for treatment with bioremediation techniques. The characterization of the following factors is essential for bioremediation.

Pollutant characterization   The composition, concentration, toxicity, bioavailability, solubility, sorption and volatilization of all pollutants.

Geohydrochemical characterization   The physical and chemical properties of the geological materials are essential to learn if the microenvironment is suitable for the biodegradative activity. The direction and velocity direction of underground flow are of fundamental interest, when the contamination has reached the water table.

Microbial characterization   It can be done by the analysis of microbial flora with respect to degradative capacity and the size of the native population with degradative potential.

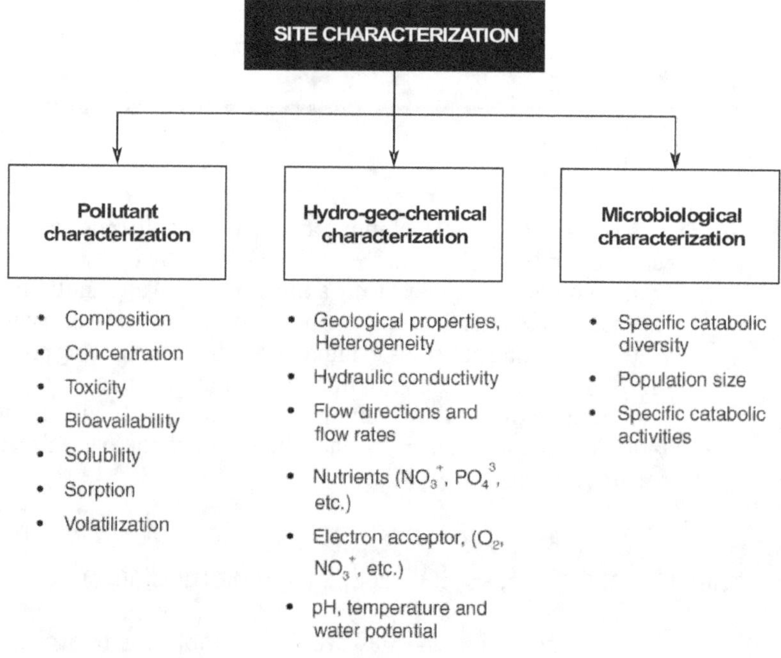

Figure 3.4 Site characterization for bioremediation

Wagner *et al.* (1986) summarized the chemical characteristics of contaminants as well as the hydro-geological characteristics of the site that are essential to evaluate the viability of bioremediation technology. The favourable and unfavourable site characteristics for the implementation of *in situ* bioremediation are given in (Table 3.3).

Strategies for the Improvement of Bioremediation Techniques

o   Addition of air or oxygen as the terminal electron acceptor for catabolic activity of microbes involved in bioremediation process.

o   Since the oxygen supplied to the *in situ* process is limited, pump-and-treat option improves biodegradation.

o   As the water solubility of alternative electron acceptors like nitrate, sulphate, etc. are high, they have the potential to improve electron-acceptor bioactivity.

Table 3.3 Favourable and unfavourable characteristics for implementation of *in situ* bioremediation

	Favourable factors	Unfavourable factors
Chemical characteristics	Small number of organic contaminants	Numerous contaminants
	Non-toxic concentration	Toxic concentration
	Diverse microbial population	Complex mixture of organic and inorganic compounds
		sparse microbial activity
	Suitable electron acceptor condition	Absence of electron acceptors
	pH 6–8	pH extreme
Hydro-geological characteristics	Granular porous media	Fractured soils
	High permeability	Low permeability
	Uniform mineralogy	Complex mineralogy
	Homogeneous medium	Heterogeneous medium
	Saturated medium	Unsaturated-saturated conditions

○ Composition of microbial communities improves bioactivity.

○ Organisms that are sensitive to water activity maximize bioactivity.

Biosurfactants in bioremediation   Surfactants are amphiphilic molecules that partition preferentially at the interface between fluid phases and alter the conditions prevailing at interfaces. The presence of both hydrophilic and lipophilic moieties is recognized within the surfactant molecule of definite molecular structure, which gives the typical properties of a surfactant (Kourkoutas and Banat, 2004). The surface-active compounds produced by microbes have attracted environmentalists because of their ecological adaptation and biological safety.

1. *Microbial surfactants*   Microbial surfactants include a wide variety of chemical structures such as glycolipids, lipopeptides, polysaccharide–protein complexes, phospholipids, fatty acids and neutral lipids. They have distinctive hydrophilic and hydrophobic moieties. Hydrophilic moiety can be ionic or non-ionic and consists of mono-, di- or polysaccharides, carboxylic acids, amino acids or peptides. Hydrophobic moieties are usually saturated, unsaturated, or hydroxylated fatty acids. Yeasts, bacteria and filamentous fungi are capable of producing surfactants. Biosurfactants include emulsifying and dispersing agents. They are advantageous because of their lower toxicity, higher biodegradability, higher selectivity and specific activity at extreme temperatures, pH and salinity. The diverse properties, structures and potential application of biosurfactants in environmental protection, crude oil recovery, health care and food processing industries, and their use in the field of agriculture and synthesis from renewable resources have made them very essential (Table 3.4).

2. *Classification of Biosurfactants*   Biosurfactants are categorized mainly by their chemical composition and their microbial origin (Figure 3.5).

3. *Applications of Biosurfactants*   Enhanced oil recovery, oil spill bioremediation/dispersion, both inland and at sea, and removal/mobilization of oil sludge from storage tanks are the most important growth areas of biosurfactants. In the case of Exxon Valdez oil spill, biosurfactants were used to emulsify hydrocarbon–water mixtures, so as to enhance the degradation of hydrocarbons in the environment.

Biomass Immobilization   One of the major problems in the use of microorganisms for the biological treatment of waste waters is the recovery from treated effluents. Immobilization technique is the best solution to solve this problem. Cell immobilization can be defined as the confinement of whole cells in an insoluble phase, which permits the free exchange of solutes from and towards the

---

Free biomass is not advantageous over the immobilized counterpart in bioremediation processes.

Table 3.4 Biosurfactants, their chemical nature and applications

Biosurfactant	Chemical nature	Organisms	Functions
Glycolipids			Lowering surface tension emulsifier
Rhamnolipids	Rhamnose + β-hydroxydecnoic acid	*Pseudomonas*	
Trachalolipids	Disaccharide + Mycolic acid	*Mycobacterium, Nocardia, Cornyebacterium, Brevibacteria*	
Sophorolipids	two β-1,2 linked glucose units + hydroxy fatty acids	*Torulopsis apicola, T.bombicola T.petrophilum*	
Lipopeptides and Lipoproteins	Gramicidin S, Subtilisin Lichenysin-A, Polymixin	B.subtilis, B.polymyxa *B.licheniformis, B.brevis*	Surface active
Phospholipids and Fatty acids		*Aspergillus, Acinetobacter, Arthrobacter. P.aeruginosa, Thiobacillus thiooxidans*	
Polymeric biosurfactants	Heterosaccharide-containing proteins (emulsan, liposan, mannoprotein)	*Acinetobacter calcoaceticus, Candida lipolytica, Saccharomyces cerevisiae*	
Particulate surfactants	Proteins, phospholipids and lipopolysaccharides (microemulsion)	*Acinetobacter, Cyanobacteria*	Alkane uptake activity, Surfactant

Aerobic microorganisms have more energy for cell synthesis compared to anaerobic microbes.

biomass but at the same time isolates the cells from their surrounding medium. Free biomass is not advantageous over the immobilized counterpart in bioremediation processes. During the biosorption process, free biomass malfunctions because pressure drops across a fixed-bed column during down-flow operations. This is due to cell clumping, and excessive hydrostatic pressure is required to generate a suitable flow-rate. Free cells are suitable only for a limited number of applications such as discontinuous reactors (Blanco *et al.*, 2003). The above constraints can be overcome by using an appropriate immobilization method for the selected biomass. Thus, biomass immobilization provides good handling and operation characteristics to biomass (Table 3.5).

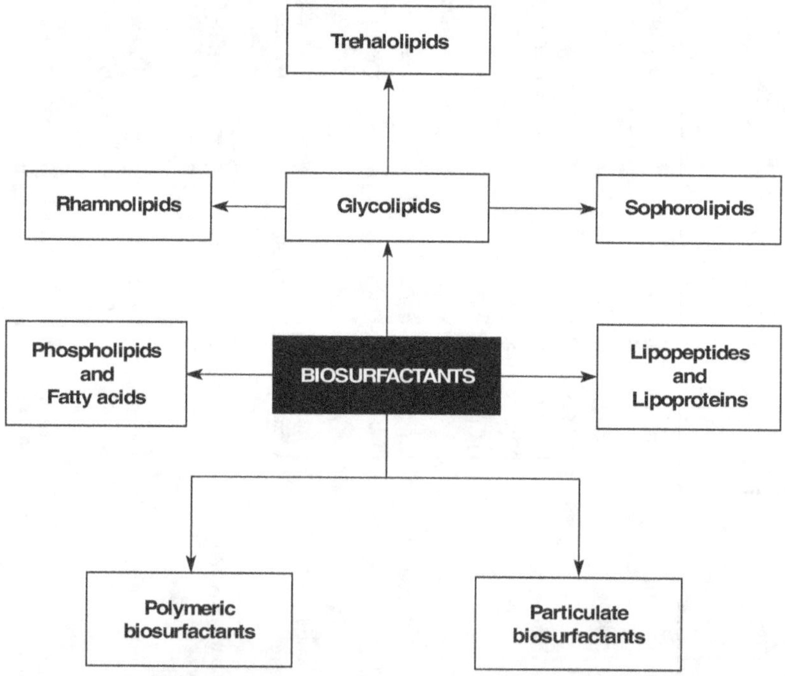

Figure 3.5 Classification of biosurfactants

Table 3.5 Microbial biomass for immobilization and their mechanism

Organisms	Enzymes	Mechanism
Clostridium thermocellum	Cellozome	Hydrolysis
Cellovibrio gilvus *Ruminococcus flavefaciens* *Clostridium thermocellum*	Cellobiose Phosphorylase	-
Cellulomonas fimi	Cellulose	Extracellular proteolysis
Pseudomonas spp.	Endoglucanase	-
Trichoderma reesei *T.koningii* *Penicillium funiculosum* *Fusarium solani* *Aspergillus fumigatus* *Sporotrichum thermophile* *Monilia*	*exo*-β-1,4-glucanase *endo*- β-1,4-glucanase β-glucosidase	Solubilization

Immobilization of biomass in a matrix that is too dense can significantly reduce contaminant loading. Aggressive immobilization leads to beads with lesser porosity and the conduct time between the immobilized biomass and the water with contaminants must be increased to allow for the increased diffusion time. Contaminants, especially metals adsorbed onto the immobilized biomass, can be eluted in the same way as in the case of free biomass, i.e., by means of the appropriate desorbing agent. Optimum eluent consists of moderately acid solution. Hydrochloric, nitric and sulphuric acids are the common eluents. Metals such as Au, Ag and Hg can be desorbed using chelating compounds such as EDTA or complexing agents such as thiourea.

1. *Substances for immobilization* A higher number of substances have been examined for the immobilization of different types of biomass. Be dell and Darnall, (1990) reported that polyacrylamide, calcium alginate and silica can be used as support for algal immobilization. Agar, agarose, κ-carrageenan and diatomaceous earth are the

natural materials used as immobilization matrices. Poly-urethane and polyvinyl foams, polyacrylamide, ceramics, epoxy resin and glass beads are the synthetic materials frequently used as biosolvent supports.

Yeasts, bacteria and filamentous fungi are capable of producing surfactants.

2. *Cell immobilization techniques*   Blanco *et al.* (2003) classified immobilization techniques as physical and chemical, depending on the type of binding involved (Figure 3.6). Physical immobilization is based on the natural tendency of cells to form flocs or to absorb onto insert surfaces. Binding to supporting surface, cross-linking and entrapment into the porous structure of a polymeric matrix are the important chemical immobilization strategies. Biomass binding to the surface occurs by adsorption, chelation, ionic bonding or covalent bonding. Gelation, polymerization and insolubilization are the different methods of entrapment.

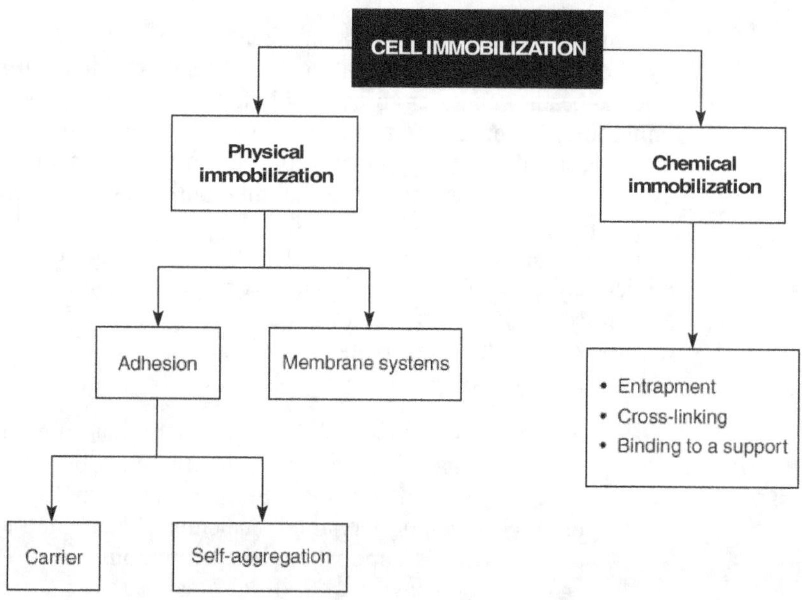

Figure 3.6 Cell immobilization methods

3. *Advantages of Immobilization* Geraats (1992) has listed the advantages of immobilization as follows:

○ Biocatalyst can be kept in reactor relatively easily.

○ High flow rates in a reactor are possible in combination with high biocatalyst retention. High flow rates can be used without loss of biocatalysts.

○ Improved physical and chemical stability. Biocatalysts can be protected from shear forces due to aeration or mixing and from the negative (inhibitory) influence of certain compounds outside the immobilization matrix.

4. *Immobilized microbes in bioremediation* Microbes could be immobilized either in a viable or non-viable form depending on its end use. The immobilized cells have been investigated for their application in pharmaceutical, food and dairy industries, in waste-water treatment, in biofuels and in the synthesis of various chemicals (Table 3.6).

Table 3.6 List of immobilized microbes in bioremediation

Organism	Support	Application
Pseudomonas putida	Agar	Degradation of caffeine
Alcaligenes denitrificans	Alginate	Degradation of *n*-valeric acid
Pseudomonas putida	Agar	Phenol degradation
Alcaligenes sp. *Enterobacter* and *Citrobacter* sp.	Polyacrylamide	Divalent metal cations, DDT
Escherichia coli	Carrageenan	Aromatic rings (xenobiotics)
Nocardia rhodochrons	Alginate/ Polyacrylamide	Cholesterol
Aspergillus sp.	Alginate	Chlorinated benzoate
Penicillium frequantans	Clay, sand	*n*-alkanes
Scenedesmus actus, S.obliqus	Carrageenan	Waste water

BIOREMEDIATION MONITORING AND CASE STUDIES

Bioremediation as a treatment technology requires the demonstration of its effectiveness, reliability, predictability and its advantages over conventional treatment technologies (Heitzer and Sayler, 1993). Interdisciplinary, systemic monitoring and evaluation strategy at each process-scale level of bioremediation is essential for successful implementation (Figure 3.7). Bioremediation process can be

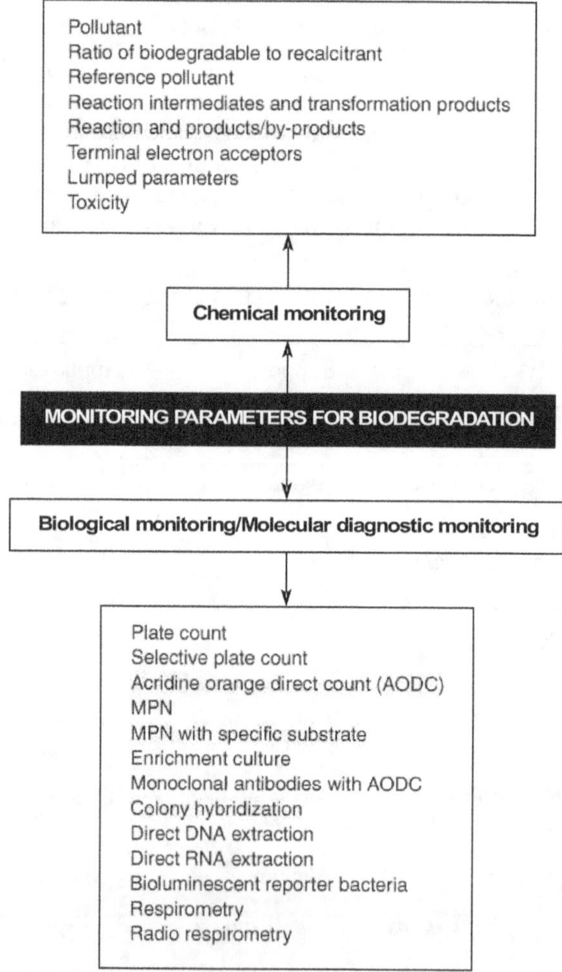

Figure 3.7 Monitoring strategies for bioremediation

subdivided into three aspects—site characterization, treatment evaluation and process scale-up. The effectiveness of bioremediation depends on five general interrelated criteria that include chemical and engineering aspects, economics, eco-toxicological aspects biological nature and aesthetic aspects.

Physical Parameters

Pollutant degradation may be incomplete due to environmental condition and microbial population. Intermediate compounds exhibit a greater mobility than the parent compounds due to aerobic hydroxylation reactions. The intermediates can be detected and characterized through rigorous extraction and combined gas chromatography/mass spectrometry (GC/MS). Intermediates can be direct indicators of *in situ* biodegradation.

Chemical Parameters

Pollutant monitoring itself is an integral part of every monitoring protocol in bioremediation. However, monitoring the disappearance of a pollutant to demonstrate its biodegradation is only useful if other monitoring parameters which demonstrate its mineralization and/or detoxification are incorporated into the monitoring and control designs (Heitzer and Sayler, 1993).

To demonstrate of *in situ* biodegradation in complex pollutant mixtures such as crude oil, ratios between easily biodegradable linear C_{17} or C_{18} *n*-alkanes and the more recalcitrant branched-chain alkanes, pristane and phytane have been used (Atlas, 1991). A decrease in the C_{18} to phytane ratio would indicate *in situ* biodegradation. This was used to demonstrate *in situ* biodegradation of crude oil on polluted beaches after Exxon Valdez oil spill. Formation of end products such as CO_2 and CH_4 provides direct proof of the complete destruction and mineralization of a pollutant. End product assay is a simple approach conducted during bioassessment and bench-scale studies. At a large process scale, gas phase CO_2 measurements are conducted for vadose-zone treatments and composting processes. Special case of end product monitoring involves the release of by-products such as chlorine during the degradation of chlorinated pollutants. Measurement of the terminal electron acceptor consumption is widely used for monitoring pollutant biodegradation at every process-scale level. At the bench-scale level, respirometric studies are useful in evaluating the efficiency of different treatments. At the field scale, gas-phase respiration measurements have been demonstrated for *in situ* vadose-zone bioremediation process (Hinchee *et al.,* 1991). Dissolved oxygen breakthrough curves can be used as indicators of ground-water bioremediation.

Biological Parameters

The impact of contaminants on the diversity of microflora can be analysed based on nucleic and fatty acid. The emergence of culture-independent techniques has allowed us to find the complexity of microbial communities and the huge, large, untapped, biotechnological resources that they represent. Direct extraction of nucleic acids (i.e., DNA, mRNA and rRNA) and fatty acids, recombinant DNA and molecular phylogeny techniques are useful to study the total microbial community structures of the soil. Toxicity and genotoxicity assays are also included in bioremediation studies. For scale-up and field implementation, composition diversity, size and relative activity of degradative microbial population is very important. This information will be essential for kinetic and modelling considerations as well as for monitoring process efficacy in stimulating specific degraders in populations. The classical plate count and most probable number (MPN) methods can be utilized for the detection of specific catabolic phenotypes.

Molecular Techniques in the Analysis of Contaminated Sites

Molecular diagnostic methods for the biological site characterization and monitoring of specific microbial populations and activities will improve the bioremediation process (Figure 3.7). Total DNA extraction from the contaminated site and subsequent hybridization with specific gene probes will avoid conventional cultivation techniques (Sayler *et al.,* 1985). Molecular biology techniques such as colony hybridizations with catabolic gene probes will be useful to detect specific catabolic genotypes. Fleming *et al.* (1993) devised a method to monitor specific *in situ* catabolic gene expression using direct isolation of mRNA from contaminated soils. Subsequent hybridization with specific gene probes allows quantification of specific catabolic mRNA. Positive correlation was observed between naphthalene dioxygenase mRNA and soluble naphthalene concentration. These data are a direct measurement of *in situ* catabolic gene expression. Use of genetically engineered catabolic bioluminescent reporter bacteria for *in situ* monitoring of catabolic gene expression and naphthalene biodegradation is a novel strategy (King *et al.*, 1992).

> *The impact of contaminants on the diversity of microflora can be analysed based on nucleic and fatty acid.*

Successful Bioremediation Stories

Oil pollution in waterways and its impacts on the environment was demonstrated by the Exxon Valdez, the largest oil spill in history, carrying 1.2 million barrel of oil when it ran into a reef off the coast of Alaska. The initial measure taken for the accident was physical washing with high-pressure water, which was not successful. However biostimulation, application of fertilizers to the polluted beaches to accelerate the growth and activities of petroleum-degrading microbes improved the conditions. Two to three weeks later, pebbles on the beaches that had been treated with fertilizers had become significantly cleaner than those in the control area (O'morchoe *et al.*, 1998).

Oil from the beaches in Prince William Sound was cleaned by bioremediation procedures. Fertilizers were sprayed on the oily beaches to enhance the growth of the oil-eating bacteria that occur in the natural environment. The fertilizer consisted of essential nutrients such as nitrogen, and phosphorous compounds mixed with emulsifying agents to solubilize the oil. The nutrients with the carbon of the petroleum provided an environment ideal for the growth of petroleum-consuming bacteria. Inipol EAP22, a mixture of oleic acid, lauryl phosphate, 2-butoxyethanol, urea and water is effective in enhancing oil biodegradation of contaminated beaches. Studies showed that nitrogen, which appears to be the limiting nutrient for hydrocarbon-metabolizing bacteria in the presence of oil spill, is believed to be essential for successful bioremediation. Oleic acid and lauryl phosphate stimulates the growth of other bacteria such as nitrogen-fixing ones, which enhance the growth of the hydrocarbon-metabolizing bacteria. The oil-eating bacteria are aerobic and the addition of nitrogen and phosphorous-containing fertilizers enhances the growth of the bacteria resulting in rapid biodegradation (Bailey *et al.,* 2005).

Another case involved the use of sulphate-reducing bacteria (SRB) in 2003 to treat contaminated ground water at the Budelco zinc-smelting works in Netherlands. The pilot plant comprised a 9-m³ stainless-steel sludge-blanket reactor using SRB and was developed by Shell Research Ltd. and Budelco B.V. This plant successfully removed toxic metals, primarily Zn and sulphate, from contaminated ground water at the long-standing smelter site by precipitation as metal sulphides (Ahmad and Zakaria, 2004)

> *Molecular diagnostic methods for the biological site characterization and monitoring of specific microbial populations and activities will improve the bioremediation process.*

Southern Petrochemical Industries Corporation (SPIC) located at Chennai, India has developed a new strategy which is different from activated sludge process (ASP). The strategy involves the isolation of targeted natural and non-pathogenic organisms, which are induced, enriched and tested for actual effluents to be treated. Urea biohydrolysis (UBH), Glycol Bio Treatment (GBT) and Phenol Biotreatment (PBT) are the successful bioremediation strategies developed by SPIC (Gupta, 2004). UBH process involves the use of ureolytic and non-ureolytic bacterial species, which can tolerate up to 28,000 ppm of urea or 30,000 ppm of ammoniacal nitrogen. GBT involves the use of mixed culture of *Pseudomonas* and *Aerobacter,* which has a synergistic effect. *Pseudomonas* converts propylene oxide (PO)/Propylene glycol (PG) into volatile acids (lactic, pyruvic and formic) and *Acetobacter* degrades the acids into CO_2 and H_2O. PBT involves the use of a mixed bacterial population, which can degrade phenol up to 2,800 ppm, ammonia up to 4,000 ppm, thiocyanates up to 1000 ppm and cyanides up to 100 ppm in a single stage. The microbial process can degrade the oil refinery waste waters that contain phenol, oil, grease, sulphides and other inorganic salts.

Bioremediation was selected as the method of choice to clean up an abandoned refinery site in southern California. The thirty-two acre site was located in a prime industrial area and the goal was to clean the site to a low-enough level so that commercial buildings could be built. The initial contamination levels for the site ranged from a low of 1500 ppm to a high of 30,000 ppm. The site was sectioned off into several treatment zones, and a bioremediation programme was begun using a consortia of microorganisms supplied by Solmar Corporation of Orange, CA. Since the site had been contaminated on and off for a period of forty years with little or no sign of decontamination by indigenous organisms, it was concluded that a bioaugmentation program could accelerate the remediation process. The treatment was conducted over a period of six months. While some areas were being treated, other areas were being taken out of service until the entire tank farm was dismantled. As areas were taken out of service, treatment was begun to remediate those sections of the property. The twenty-nine acres of the area was certified as clean within a period of one year. The balance, which has been used as the dumping area, is still being remediated.

In the city of Carson, California, a particular site was condemned since it had been used as a petrochemical tank storage site and salvage operation. Rather than sealing the contaminants at the site under buildings and parking lot, it was decided to get rid of the contaminants. The site had been earmarked as a park; and the city officials were concerned that, if the contaminants were left in place, they may endanger the health of the children using the park (Stein, 1987). The price for hauling away the contaminated soil for proper disposal was estimated to be $2 million. The estimated amount of contaminated soil was approximately 10,000 cubic yards. A bioaugmentation program was proposed and adopted at the site. The cost of the clean up was less than $132,000, and the city began seeking bids for its most elaborate recreation facility.

When the Sacramento Utilities District purchased a small parcel of land to expand their existing parking lot, they were unaware that the land had been previously contaminated with diesel fuel. Once the contamination had been detected, the Utilities District decided to take it upon themselves to clean up the site. The District realized that merely excavating and hauling the contaminated soil to a dump site was just transferring the problem to another site. In keeping with the Districts policy of concern with the environment, other alternatives to land disposal were sought. Upon examination of treatment options, the District decided to implement a bioremediation programme using bioaugmentation as the source of organisms. The bioremediation of the 2000 cubic yards of contaminated soil reduced the Total Petroleum Hydrocarbon levels from 2800 ppm to less than

38 ppm in approximately 74 treatment days (Rittenhouse, 1998). The cost of treatment was $360,000 less than the total price of disposal without the inherent liability.

Bioremediation was the method of treatment opted for to treat 1500 cubic yards of diesel-contaminated soil at the former King's Truck Stop in Sacramento, CA. The project reduced the diesel contaminant levels from 3000 ppm to less than 30 ppm in approximately 62 treatment days.

In situ bioremediation was necessary to clean up contamination from a ruptured transfer line that passed under a railroad track. A jumbo tank car had been moving on the track as solvents were being pumped through the line. The resulting rupture led to a loss of 300 to 400 gallons of solvent at a depth of 38 inches beneath the surface along 120 feet of the track. A continuously recirculating ground injection system was designed and installed to treat the contaminated soil.

BIODEGRADATION PROCESS

4

COMMERCIAL BIOAUGMENTATION PRODUCTS

Bacterial cells such as *Alcaligenes, Acinetobacter, Arthrobacter, Beijerinkia, Klebsiella, Flavobacterium* and *Pseudomonas* contain specific plasmids which encode enzymes for degradation of specific contaminants. Some of the strains may be irradiated to enhance their ability and mutants are selected. They are tested in the laboratory for biodegradation ability and the selected strains are used in large fermenters to get a mass culture. Commercial bioaugmentation products are single cultures or consortia of microbes with certain degradative properties or their other desirable characteristics. At present they are used in industrial waste-water treatment plants. The use of added microbes for treating hazardous wastes such as phenol, ethylene glycol and formaldehyde has been attempted. Bioaugmentation with parachlorophenol-degrading bacteria decomposed 96% parachlorophenol in 6 hours. Cells of *Candida tropicalis* have been used for removal of high concentration of phenol present in freshwater. Bioreactor ability to dechlorinate 3-chlorobenzoate was increased after the addition of *Desulfomonile tiedijei* to a methanogenic upflow anaerobic granular sludge blanket.

Dubey and Maheshwari (2005)

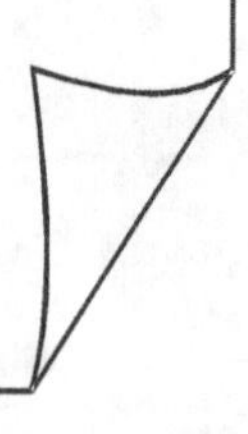

Global deterioration of the environmental quality is due to the large-scale production of a variety of chemical compounds that are different in structure. The chemical compounds are highly toxic and resistant to biodegradation. Approximately 6×10^6 chemical compounds have been synthesized, with 1,000 new chemicals being synthesized annually. Almost 60,000 to 95,000 chemicals are in commercial use (Kannaiyan, 2001). Technologies using microbes with extensive biodegradative capacities have been developed to treat the introduction of synthetic chemicals into the environment by deliberate or accidental means. Biodegradation of organic compounds has been used for at least a century to treat sewage, solid waste, industrial residues and a series of man-made xenobiotics. However some of the contaminants are recalcitrant to biodegradation due to several reasons.

Biodegradation can be defined as the decomposition or destruction of contaminant molecules by the action of the enzymatic machinery of a biological system. Degradation of the organic type of contaminants by microbes leads to complete mineralization by releasing carbon dioxide. Biodegradation of organic compounds by soil microbes involves a process known as mineralization whereby microbes convert the organic molecules to carbon and energy for growth and multiplication, releasing the inorganic forms of N, P, S or other elements. In this process, the parent molecule becomes detoxified or rendered harmless to life by enzymatic reaction. Presence of co-substrates might be required to support the energy-yielding activity. Biotransformation or bioconversion of contaminants makes them less toxic entities. Inorganic contaminants can be transformed into different entities or retained by the cells without degradation. Inorganic contaminant reduction can be observed only when the microbial activity occurs in water where compounds move from the aqueous phase to the inside of the cell.

Agents for Biodegradation

Tiny microbes such as bacteria, algae, fungi, etc. play a vital role in the process of decomposition. This is due to their catabolic versatility, the diversity in species composition and their ability to perform reactions at a much faster rate per unit weight as compared to other organisms. The microbes perform specific biochemical reactions which degrade macromolecules into smaller units in a step-wise fashion. Usually a series of organisms are required to bring about the complete degradation of large molecules or a group of molecules. Introduction of biopreparations into polluted sites (bioaugmentation) may enhance remediation owing to the degradative ability of added microbes. Transfer of metabolic capacities to indigenous bacteria by conjugation also increases the degradative microbes essential for bioremediation (Binulal *et al.,* 2003). It may be the most efficient strategy to allow natural plasmids from the introduced strains to spread to indigenous populations that are already established and adapted to their niches. Microbial biodegradation occurs due to the production of certain enzymes which act as catalysts for transformation reactions induced by the energy-yielding process. Energy is generated during breakdown of complex organic substances by heterotrophs, oxidation of inorganic compounds by autotrophs and photosynthetic reactions of phototrophs (Jogdand, 2004). Electron donor and acceptor, micro-environmental conditions for catabolic enzyme production and expression are essential for the occurrence of transformation reactions. In heterotrophic microbes, the electron donor used as a source of carbon and energy will be the organic contaminant. It supplies energy required by metabolism through electron transfer during the oxidation–reduction reaction to the electron acceptors.

Enzymes for Biodegradation

In general, the degradative pathways that have evolved towards the catabolism of nitro-substituted aromatic rings, are catalysed by common enzyme types, i.e., mono- and dioxygenases, and dehydrogenases produced by common microbial groups belonging to *Pseudomonas, Nocardia* and *Arthrobacter.* Enzymes are the catalytic units responsible for different molecular pathways in microbes. They alter the speed of reaction without altering themselves and they may be extracellular or intracellular. Extracellular enzymes break down the large complex molecules and enable their entry into the cell. The genes of certain enzymes have inducible operons while others are constitutive. Enzymes which carry on different reactions in microbial cells are responsible for metabolic pathways, which form a base for energy generation and cell synthesis.

Optimal Conditions for Biodegradation

One of the basic necessities of biodegradation is the presence of the microbe which can perform the task of decomposing specific contaminants. In many cases, a number of microbial species have to act one after the other—each performing its own specific function—to bring about the complete degradation. Contact between the microbe and contaminant should be established for biodegradation. Normally pollutants induce the microbe to produce enzymes essential for degradation. There is a lag period during which microbes adjust to the presence of the organic compound, which may extend to several days or weeks. After the lag phase, depending upon the nature of the compound, detoxification or partial detoxification can occur by any one of the several processes such as hydrolysis, dehalogenation, deamination, methylation, nitro reduction, deamination, cleavage of ether linkages, conversion of nitrite to amine and conjugation (Subba Rao, 2000). Favourable environmental factors such as water, suitable pH, temperature, etc. will also speed up biodegradation.

> *Transfer of metabolic capacities to indigenous bacteria by conjugation also increases the degradative microbes essential for bioremediation.*

Nature of Biodegradation Reactions

Biodegradation reactions include oxidation, reduction, hydrolysis and structural rearrangements. Simple organic compounds such as carbohydrates, proteins, fats and nucleic acids are quickly degraded. Molecules with branched chains, increased degree of substitution, simple or fused aromatic rings, cyclo-paraffins, etc. will be degraded with difficulty. Degradation of these types of molecules involves step-wise removal of substituted groups or atoms. The lipophilic compounds that pose greater hazard to the biological system are converted to water-soluble or hydrophilic compounds. Due to this reaction, the movement within the biological system will be restricted and in the course of time it will be excreted. Conjugation with glucuronic acid, sulphates, amino acids, acetylation and methylation are some of the common reactions which transform a contaminant molecule into a water-soluble compound.

> *Biodegradation reactions include oxidation, reduction, hydrolysis and structural rearrangements.*

Dynamism of Biodegradation Reaction

The process of biodegradation has the capacity to adjust its activity according to the amount of contaminant. The rate of biodegradation depends on the ratio between the concentration required to saturate the system and the concentration of the contaminant within a biological system. Below saturation, degradation reactions decompose a constant fraction of the total content of the contaminant per unit time (Figure 4.1). Larger loads of contaminants are decomposed at a faster rate which slows down as the concentration of contaminants within the system diminishes. The term "half-life" is used to measure the period of retention of contaminants or any foreign substance within a biological system. Half-life of a foreign substance in a biological system denotes the duration of time within which its concentration is reduced to half its original value.

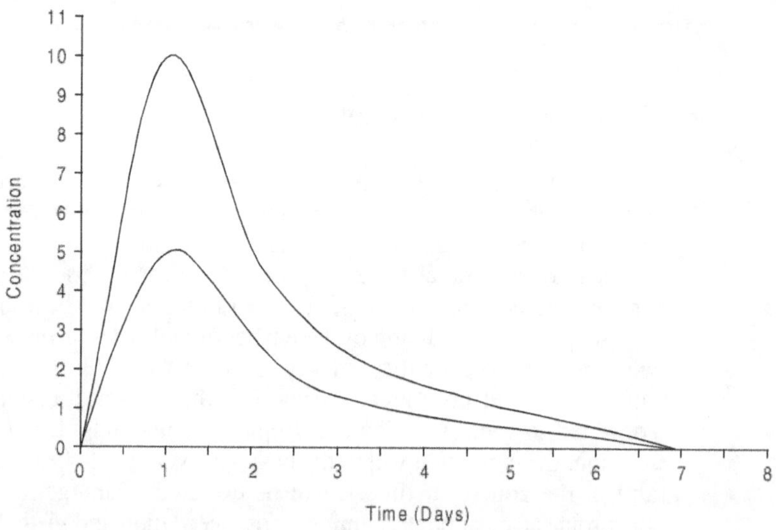

Figure 4.1 Biodegradation of contaminants in a biological system

Biological systems are capable of decomposing small quantities of contaminants efficiently. The contaminants which were not decomposed completely before the entry of fresh dose, tend to accumulate within the system. The concentration of foreign substance and the rate of degradation or elimination are directly proportional to each other. This results in an equilibrium between the intake and rate of decomposition, which adjusts itself in such a way as to keep the concentrations of the contaminant at reasonably constant level. This can be demonstrated by the following example (Figure 4.2).

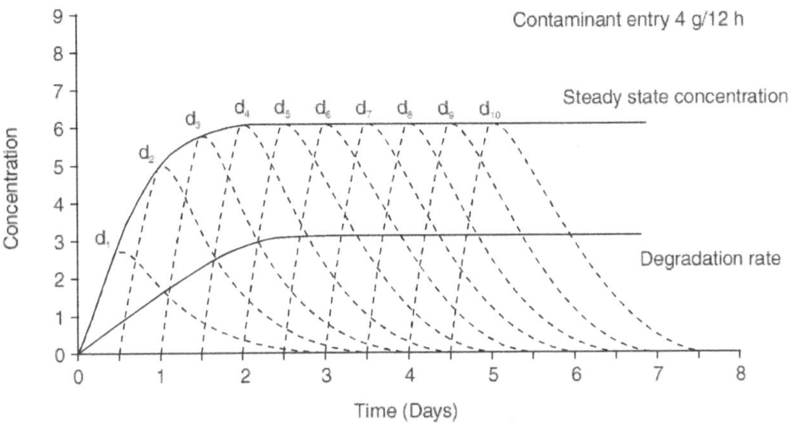

Figure 4.2 Steady state concentration of contaminants

A contaminant which has a half-life of 12 hours is introduced in a biological system at a rate of 4 after every twelve hours (d_1, d_2, d_3 ......). A concentration of 4 g per litre shall be attained in the system after 12 hours when the second dose of 4 g is introduced. Samples collected after 24 hours show a concentration of about 6 g only. This is because in the preceding 12 hours (the half-life) the initial concentration was reduced to 2 g to which the next dose of 4 g was added. The third dose will raise the concentration to about 7 g only, since by the time it reaches the system, the earlier concentration of 6 g would be reduced to 3 mg per litre. Similarly 7 g will be reduced to 3.5 g after 12 hours when the next dose of 4 g enters the system. This dose raises the overall concentration to 7.5 g only. Each dose entering the system is added to half of the total content of the contaminant that was present 12 hours earlier. The toxic material attains a concentration of 4 g per litre after the first dose. The second dose raises it by 2 g, the third by 1 g, the fourth by 0.5 g, the fifth by 0.25 g, the sixth by 0.125 g and so on. Finally a steady state level is attained. The rate of biodegradation reaction rises after a few doses to prevent any further increase in concentration. The input equals the output (Asthana and Asthana, 2001).

> *The microbes perform specific biochemical reactions which degrade macromolecules into smaller units in a step-wise fashion.*

State of Contaminants with Different Half-lives

Half-life and the rate of uptake of contaminant are the important factors that affect the degrading biological systems. Substances which have a longer half-life tend to attain a higher concentration if the rate of input is same (Figure 4.3). In the natural environment, the biological system usually receives small exposures, which are spread over long durations. A constant rate of biodegradation could have caused simple arithmetic addition of contaminants. As the time increases, the contaminant will reach a lethal level. However, the biodegrading organisms are capable of adjusting themselves to the rate of flow, whatever be the duration of exposure. When the mechanism fails, even a very meagre amount of contaminant disturbs the biological system and the contamination increases.

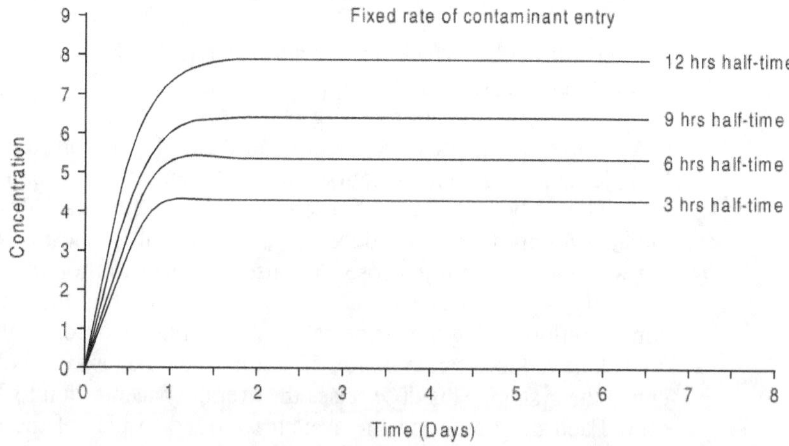

Figure 4.3 Steady state of concentration attained by similar doses of contaminants having different half-lives in a living system

> *Biological systems are capable of decomposing small quantities of contaminants efficiently.*

Rate of Contaminant on Biodegrading Organisms

Biodegradation reactions are forced to occur at their maximum pace to prevent increase in contaminant concentration. The system gets strained to work at its possible rate at its critical concentrations. Self-purification is possible below the critical level. Any further rise in the concentration beyond the critical level will affect the biodegradation machinery and its self-purifying capacity is lost. At this enfeebled pace of decomposition, biologically unlimited contamination sets in and the recovery is not possible.

> *Enzymes are the catalytic units responsible for different molecular pathways in microbes.*

BIODEGRADATION OF XENOBIOTICS

Xenobiotics are the chemicals synthesized by humans that have no close counterparts. The term "Xenobiotic" (stranger to life) is derived from the Greek word *Xenon* meaning strange and *bios* meaning life. Xenobiotic usually implies something foreign to the biosphere. Natural biodegradation of xenobiotics is often a slow process and depends upon the complexity, concentration, toxic intermediates, pH, temperature, oxygen availability, redox potential and the type of microbes. Microbial degradation of the xenobiotics is difficult due to the following reasons.

1. They are persistent in the environment since the transport mechanisms and metabolic pathways for them have not been evolved in microbes.

2. The xenobiotics have molecular structure and chemical bond sequences not recognized by the existing degrading enzymes.

3. The unusual bonds and bond sequences such as in tertiary and quaternary carbon atom are highly condensed aromatic

rings and excessive large molecular size in the case of plastics and polyethylene.

4.   The lack of some essential growth factor made it unable for the naturally occurring microbes to metabolize these compounds.

In most of the cases, a mixed consortium in a sequential mechanism rather than individual species, brings about xenobiotic degradation. The existence of a consortium of species enhances the degradative process by supplementing nutrient and cofactor needs of each other in the community. Different microbes carry out difficult steps in the degradation. For example, the degradation of 3-chlorobenzoate is performed by three different groups of bacteria.

The hydrogen generated in step 2 is utilized by the group of microbes doing the reductive halogenation in step1 and the process is outlined in the reaction shown in Figure 4.4.

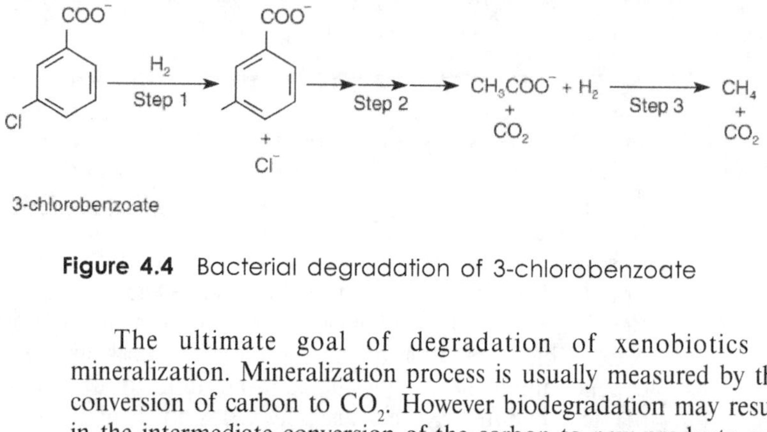

3-chlorobenzoate

Figure 4.4 Bacterial degradation of 3-chlorobenzoate

The ultimate goal of degradation of xenobiotics is mineralization. Mineralization process is usually measured by the conversion of carbon to CO_2. However biodegradation may result in the intermediate conversion of the carbon to new products, not achieving mineralization. These biological products will eventually be degraded to CO_2. For example, the acetate formed in the degradation of 3-chlorobenzoate is readily metabolized by both anaerobic and aerobic microbes. The crucial step in the degradation of 3-chlorobenzoate is the reductive elimination of chloride ion, a reaction that is carried out by relatively few microbes. Anaerobes and aerobes degrade xenobiotics. Anaerobes often use reductive processes to degrade organics, and aerobes use oxidative processes. Reductive, oxidative and hydrolytic degradation are the mechanisms involved in xenobiotic biodegradation (Bailey *et al.,* 2005).

Reductive Degradation

Bacterially mediated reductive dehalogenation is an important pathway for the destruction of unreactive haloorganics. The reaction usually occurs in anaerobic environments such as marine sediments and landfills with the replacement of halogens with hydrogen. The principal pathway for the reductive dechlorination of 2,3-dichlorobenzoate is as shown in Figure 4.5.

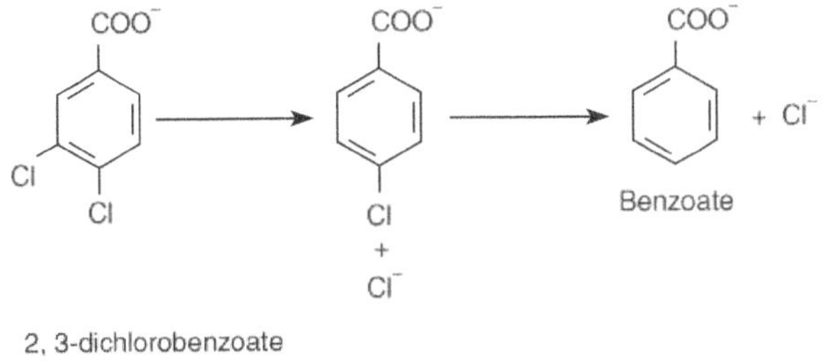

Figure 4.5   Pathways of reductive dechlorination of 2,3-dichlorobenzoate

Reductive dechlorination is the first and important step in the degradation of heavily chlorinated organics like pentachlorophenol (PCP). PCP is rapidly converted to tetrachlorophenols, but the subsequent loss of chloride in the process of forming tri-, di- and monochlorophenols proceeds much more slowly than the initial loss of chloride ion.

Oxidative Degradation

The energy derived from the oxidation of xenobiotics with atmospheric oxygen drives the process of oxidative degradation. These oxidations occur with lightly chlorinated chloroorganics. The overall oxidative degradation of chlorobenzene is shown in (Figure 4.6).

Microbial deoxygenase enzyme catalyses the formation of a cis-dihydrodiol, which is then oxidized to a halogenated catechol. Enzyme-catalysed oxidative ring cleavage takes place either between the two phenol groups or adjacent to one of the phenol groups to give dicarboxylic acids, which are readily degraded to simpler compounds and eventually to CO_2, H_2O and Cl^-.

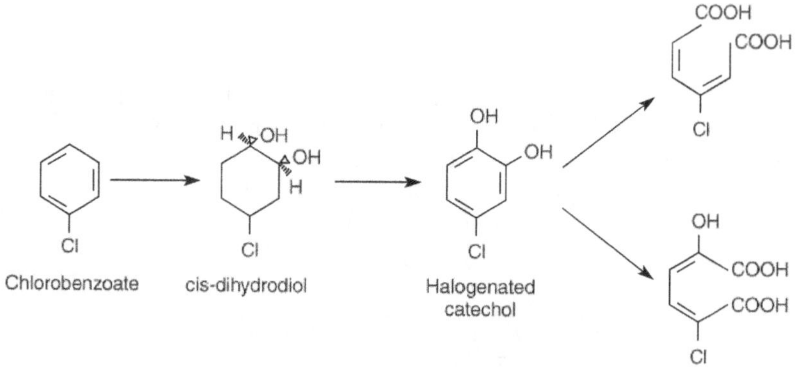

Figure 4.6 Oxidative degradation of chlorobenzenes

Hydrolytic Degradation

Aerobic and anaerobic microbes catalyse hydrolytic cleavage of halogens, esters, amides, and other groups. Dichloromethane (methylene chloride), a compound that decomposes very slowly in the presence of water, is hydrolysed by microbes to formaldehyde. The enzyme-catalysed reactions usually proceed by the SN_2 displacement of chloride ion by the thiol peptide, glutathione (G-SH). The resulting intermediate is hydrolysed to glutathione and formaldehyde. The microbe that catalyses the reaction utilizes formaldehyde as an energy source (Figure 4.7).

$$G{-}SH + CH_2Cl_2 \longrightarrow \underset{\displaystyle HCl}{\underset{+}{G{-}SCH_2Cl}} \longrightarrow \underset{\displaystyle HCl}{\underset{+}{G{-}S{=}CH_2}} \xrightarrow{H_2O} G{-}SH + CH_2O + H^+$$

Figure 4.7 Hydrolytic cleavage of halogens

HYDROCARBONS (HCs) AND THEIR DERIVATIVES

Hydrocarbons are the organic compounds that contain only carbon and hydrogen. Most of the chemicals in gasoline and other petroleum products are hydrocarbons, and are divided into two categories: **aliphatic** and **aromatic**. Aliphatic hydrocarbons include **alkanes,**

alkenes and **alkynes**. The alkanes are saturated hydrocarbons (methane) and are fairly inert. The alkenes, generally called olefins, are unsaturated and highly reactive. The alkenes (such as ethylene), in the presence of sunlight, react with nitrogen dioxide at high concentration to form secondary pollutants such as PAN (peroxyacetyl nitrate) and ozone. The alkynes are highly reactive and relatively rare. Aromatic hydrocarbons are derived from or related to **benzene** and are biochemically and biologically active. Aromatics do not display the reactivity characteristics of unsaturated aliphatic hydrocarbons, but the polynuclear groups of aromatic hydrocarbons are carcinogenic.

> *A mixed consortium in a sequential mechanism rather than individual species, brings about xenobiotic degradation.*

Most of the natural hydrocarbons are from biological sources, though some amounts come from geothermal areas, coalfields, neutral gas from petroleum fields and natural fires. Hydrocarbons are considered to be the parent compounds for organic materials. Crude oil is the predominant source of the hydrocarbon compounds used for combustion and industrial processes. It is estimated that 2.1 million metric tons of crude oil and hydrocarbons enter the environment each year (Bailey *et al.*, 2005). The persistence of anthropogenic aromatics in the environment has caused great concern due to their toxicity, mutagenicity and bioconcentration in higher organisms. Aromatic hydrocarbons and their derivatives are used in industrial operations such as chemical solvent manufacturing, pesticide production and other petroleum-based industries. The use of microbes to degrade aromatic pollution is a low-cost approach for the treatment of industrial effluents, contaminated sediments, soil and ground water.

Aliphatic Hydrocarbons

The *n*-alkane oxidation that involves oxidation of the terminal methyl group to form a primary alcohol using molecular oxygen is the main component of aliphatic hydrocarbon degradation. Two enzymatic systems have been elucidated. The first involves cytochrome P-450, an iron protein and flavoprotein that are sensitive to carbon monoxide. It is found in *Cornyebacterium* sp. and in some *Candida* strains. Cytochrome P-450 is a super family of haemoproteins, which are found in bacteria, fungi, animals and plants. The second system found in *Pseudomonas* group is a flavoprotein and rubredoxin components for electron transfer in catalytic system. In bacteria, components essential for enzyme synthesis are encoded in the plasmids. Some microbes are able to start *n*-alkane oxidation in a non-terminal site, forming ketones. This oxidation may also be catalysed by mono-oxygenases. The ketones are transformed to esters by insertion of oxygen and then hydroxylated to form one acid and alcohol.

> *The ultimate goal of degradation of xenobiotics is mineralization.*

The degradation of the primary alcohol is carried out by the alcohol–aldehyde dehydrogenase pathway to obtain the corresponding fatty acid. These enzymes usually have a bound NAD, and are induced by the presence of hydrocarbons. Generally, the fatty acid is transformed to the acyl-CoA derivative, and then cleaved to form acetyl CoA. Many microbes are able to make oxidations to produce α-hydroxyl fatty acids, which are decarboxylated later (VazQuez-Duhalt, 2003).

Mechanisms for oxidation of n-alkanes to primary alcohols Terminal alkanes are attacked, in most cases at both ends of the molecule, producing a variety of products (Figure 4.8). Internal alkanes are modified by the addition of a water molecule across the double bond. Secondary alcohols or ketones are transformed to esters, and then hydrolysed to form the alcohol and carboxylic acid. Isoprenoids and cycloalkanes are degraded by *Pseudomonas citronellolis* and *Nocardia* sp. respectively.

Aromatic Compounds

Enormous research work has been carried out on the mechanisms of hydrocarbon degradation by microbial cells. Aromatic compound degradation is of major importance because of their widespread occurrence in natural as well as synthetic materials. Microbes facilitate the re-entry of the carbon from these aromatic compounds into the natural biological cycles, preventing their accumulation in the environment. Herbicides, pesticides and many industrial products

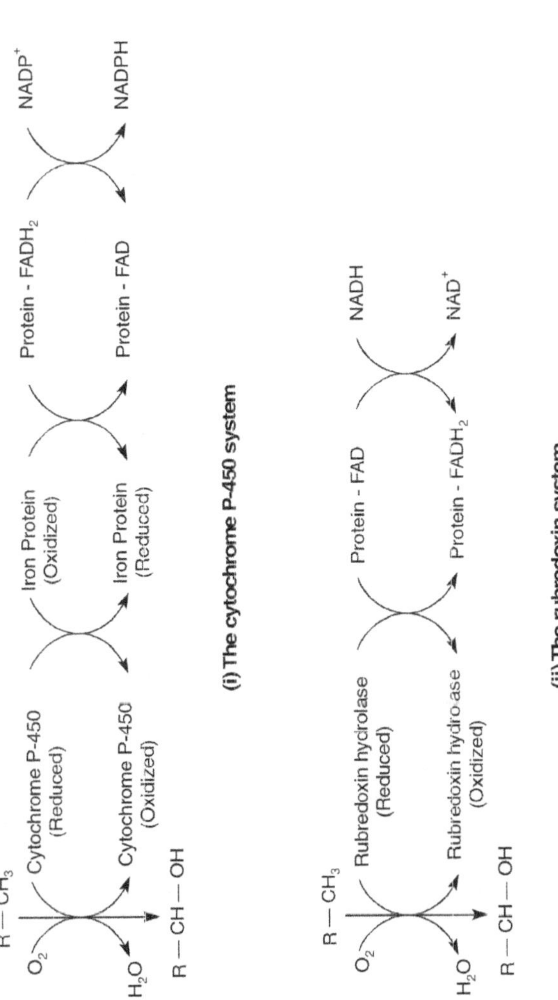

Figure 4.8 Mechanisms for oxidation of n-alkanes to primary alcohols

> *Herbicides, pesticides and many industrial products that contain aromatic compounds contaminate ground water and get accumulated in plants/animals leading to serious toxicological problems.*

that contain aromatic compounds contaminate ground water and get accumulated in plants/animals leading to serious toxicological problems. Aromatic hydrocarbon degradation is performed by a large number of microbes and several mechanisms are available. Microbial degradation of aromatic compounds is done by ring fission, which is predominantly an aerobic process. Microbes degrade these compounds either totally or partially depending on the number of aromatic units and especially on the type of substituents (Jothimani *et al.*, 2003). The basic reaction sequences involved are:

1. transformation of side chains

2. formation of ring fission substrates

3. ring cleavage and transformation of the products of ring fission into common compounds of intermediary metabolism

Presence of molecular oxygen determines anaerobic or aerobic biodegradation of aromatic compounds. Excess molecular oxygen in the environment compared to organic carbon is responsible for aerobic organic matter degradation. However, anaerobic degradation predominates due to low solubility and slow mass transport of oxygen in water.

Mechanism of degradation of aromatic compounds   Aromatic ring compound degradation has been studied in a wide variety of bacteria and fungi including *Acinetobacter, Alcaligens, Agrobacterium, Azotobacter, Bacillus, Bradyrhizobium, Rhizobium, Moraxella, Nocardia, Neurospora, Aspergillus, Trichosporon, Candida, Rhodotorula* and others. However an intensive study has been made with various species of *Pseudomonas*. The fundamental aspect of the metabolism of aromatic ring compounds is that they are converted to rings containing adjacent hydroxyl groups that facilitate ring cleavage. Several mechanisms of ring oxidation are known (Moat and Foster, 1988). Most yeasts and filamentous fungi metabolize aromatic compounds through reactions of the β-ketoadipate pathway. *Trichosporon cutaneum* cannot cleave the protocatechuate ring but oxidizes it to 1,3,4-trihydroxy benzene, which is cleaved to methylacetate. *Rhodotorula graminis* metabolizes aromatic compounds by β-ketoadipate pathway. Enzymes of β-ketoadipate pathway are inducible in *Rhizobium* and *Agrobacterium* and constitutive in *Bradyrhizobium*.

Catechol is the reduced intermediary formed by ring fission during aerobic metabolism of microbes. Catechol is converted to pyruvate or acetaldehyde and the end products are catabolized through tricarboxylic acid (TCA) cycle. Ortho-fission, meta-fission and gentisic acid pathways are the mechanisms by which aerobic cleavage of aromatic ring is achieved through dioxygenases (Figure 4.9). Ortho-cleavage produces muconic acid by an intradiol dioxygenase (catechol-1, 2-dioxygenase), and meta-cleavage to form hydroxymuconic aldehyde by an estradiol dioxygenase (catechol-2, 3-dioxygenase).

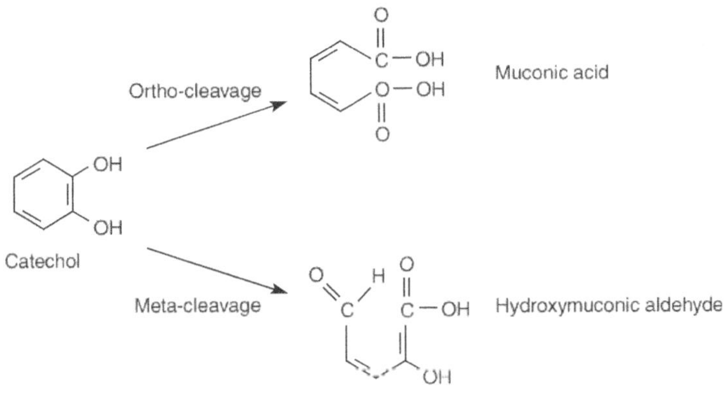

Figure 4.9 Aromatic ring cleavage

Protocatechuric acid is another substrate for the aromatic ring cleavage during the biodegradation of aromatic compounds. The enzyme involved in ring cleavage is protocatechuric acid-2, 3-dihydroxygenase. The ortho-cleavage between the hydroxyl groups produces β-carboxylated muconic acid. Ortho-cleavage is also known as endocleavage, while the meta-cleavage is known as exocleavage. Dioxygenase enzyme contains iron which is the active site of the molecule. Iron can be extracted by chelating compounds in the presence of reducing agents. Ironless enzyme is inactive, and can often be regenerated in the presence of ferric iron and an oxidizing agent (VaZQuez-Duhalt, 2003).

Metabolic process involved in the organisms

1. *Aerobic process* Molecular oxygen acts as the terminal electron acceptor during aerobic respiration. It is also essential for inserting the compounds during hydroxylation and ring cleavage. Oxygenase enzymes mediate these reactions. The microbial degradation of aromatic compounds are done by ring fission which is predominantly an aerobic process. The microbes degrade the compounds either totally or partially depending on the number of aromatic units and the type of substituents. During aerobic biodegradation process molecular oxygen is involved in two ways:

i. oxygen is incorporated in the aromatic ring by means of mono- and dioxygenases prior to ring fission (Chakrabarty, 1996).

ii. oxygen serves as the terminal electron acceptor of the reducing equivalents generated during the oxidation of the aromatic molecule.

2. *Anaerobic process* Aromatic compounds, in the absence of aerobic biodegradation in soil, reach the sediments where anaerobic condition prevails. Microbially mediated transformation is known to occur in the absence of molecular oxygen. Higher chlorinated aromatic compounds are degraded better under anaerobic conditions. Aromatic compounds such as benzene, toluene, naphthalene, etc. without functional groups cannot be degraded anaerobically. However chlorinated aromatic compounds such as hexachloro-benzene can be degraded anaerobically. Compounds such as sulphate, nitrate, sulphur, oxidized metal ions (Fe^{2+} and Mn^{4+}), protons and bicarbonate ions can act as terminal electron acceptors in the absence of O_2. Proton, and bicarbonate ions are primary electron acceptors employed in methanogenic environment. Phototropic bacteria can also use aromatic compounds for carbon and electron requirement in the absence of O_2 if light is available. In this case, the degradation product is a biomass rather than carbon dioxide (Evans and Fuchs, 1988).

a. *Energetics of anaerobic process*

Lack of oxygen in the anaerobic environment leads to the use of certain potential oxidants, which can be reduced by a biological system. The organisms can harness the energy, and neither the oxidation nor its reduced product is toxic at the required concentration (Moodle and Ingledeu, 1990). In the anaerobic environment, CO_2 or CO_3^-, NO_3^- or NO_2^- and SO_4^- can be used as alternative electron acceptors depending on the respiratory capabilities of the surviving bacteria. The maximum energy gain occurs in aerobic conditions followed by ferric and nitrate as respiratory oxidants. In the sulphur-reducing and methanogenic environment, the energy gain is very low.

b. *Electron acceptors is anaerobic process*

The nature of electron availability or hydrogen sink dominates the metabolism in the absence of oxygen. Nitrate, ferric iron, sulphate and CO_2 serve as the preferred electron acceptors for denitrifying, iron-reducing, sulphate-reducing and methanogenic bacteria (Table 4.1).

Table 4.1 Terminal electron accepting processes (TEAPs)

Electron acceptors	TEAPs	Environment
O_2	Aerobic	Sediment/ground water
NO_3	Denitrifying	Nitrate-rich ground water
$Fe(III)$	Iron reducing	Ground water
SO_4^{2-}	Sulphate reducing	Marine systems
CO_2	Methanogenic	Organic rich systems

Anaerobes often use reductive processes to degrade organics, and aerobes use oxidative processes.

Oxidized groups of humic acid substances can serve as electron acceptors. Such groups act as redox shuttle between immobile ferric iron and bacteria in suspension. Recalcitrant substrates such as aromatics cannot be activated in the absence of oxygen. Oxygen acts as the electron acceptor and reactant in oxygenase reaction during metabolism. The principal steps in ring cleavage of aromatics in the absence of oxygen involve activation reactions, channelling reactions to central intermediates and ring cleavage reactions. Reduction of aromatic ring to an alicyclic ring, cyclohexanone and hydrolytic cleavage are the three main reactions involved in the production of aliphatic acids, leading to common metabolic end products (CO_2, CH_4 and biomass).

Polycyclic Aromatic Hydrocarbons

Polycyclic aromatic hydrocarbons (PAHs), also known as polynuclear aromatic hydrocarbons (PNAs) and polycyclic organic matter (POM), are composed of hydrogen and carbon arranged in the form of two or more fused benzene rings in linear, angular, or cluster arrangements, which may or may not have substituted groups attached to one or more rings (Sims and Overcash, 1983). In some cases, the newly defined substituted PAH has strikingly greater toxicological effects than does the parent compound (Sirota and Uthe, 1981). The nomenclature of PAH compounds has been ambiguous in the past due to different peripheral numbering systems. Physical and chemical characteristics of PAHs generally vary with molecular weight. With increasing molecular weight, aqueous solubility decreases, and melting point, boiling point, and the log K_{ow} (octanol/water partition coefficient) increases, suggesting increased solubility in fats, a decrease in resistance to oxidation and reduction, and a decrease in vapour pressure. Accordingly, PAHs of different molecular weights vary substantially in their behaviour and distribution in the environment and in their biological effects.

Effects of polycyclic aromatic hydrocarbons (PAHs) Polycyclic aromatic hydrocarbons (PAHs) are the most potent carcinogens, producing tumours in organisms through single exposures to microgram quantities. PAHs act both at the site of application and at organs distant to the site of absorption. The evidence implicating PAHs as inducers of cancerous and precancerous lesions is becoming overwhelming, and this class of substances is probably a major contributor to the recent increase in cancer rates reported for industrialized nations. PAHs were the first compounds known to be associated with carcinogenesis (Lee and Grant, 1981). There are thousands of PAH compounds, each differing in the number and position of aromatic rings, and in the position of substituents on the basic ring system. Environmental concern has focused on PAHs that range in molecular weight from 128.16 (naphthalene, 2-ring structure) to 300.36 (coronene, 7-ring structure). Unsubstituted lower molecular weight PAH compounds, containing 2 or 3 rings, exhibit significant acute toxicity and other adverse effects to some organisms, but are non-carcinogenic; the higher molecular weight PAHs containing 4 to 7 rings are significantly less toxic, but many of the 4 to 7-ring compounds are demonstrably carcinogenic, mutagenic, or teratogenic to a wide variety of organisms, including fish and other aquatic life, amphibians, birds, and mammals. In general, PAHs show little tendency to biomagnify in food chains, despite their high lipid solubility, probably because most PAHs are rapidly metabolized. Inter- and intraspecies responses to individual PAHs are quite variable, and are significantly modified by many inorganic and organic compounds, including other PAHs.

Oxidative metabolism of polynuclear aromatic hydrocarbons (PAH) is accomplished through highly electrophilic oxidases, some of them may bind covalently to macromolecules such as DNA, RNA and proteins and induce mutations. PAHs are active inducers of liver enzymes of fish and produce metabolites which accumulate and induce cancer. Naphthalene, even in small concentrations, is capable of inducing hepatic histopathological alterations which may cause severe physio-metabolic dysfunction leading to death. The PAH-contaminated sediments are also a concern for human health because of their potential to contaminate seafood consumed by man. Fish and shellfish frequently accumulate PAH to concentrations that are several orders of magnitude greater than their aqueous solubility. Acute effects due to their exposure range from skin and lung irritation to cyanosis.

Sources of PAHs   PAHs are widely-distributed in the environment, almost ubiquitous, and have been detected in animal and plant tissues, sediments, soils, air, surface water, drinking water, industrial effluents, ambient river water, well water, and ground water (EPA 1980). About 43,000 metric tons of PAHs are discharged into the atmosphere each year, and another 230,000 tons enter aquatic environments. PAHs are released into the environment by motor emissions, volcanic activity, power and heat generation, petroleum spillage, forest fires, microbial synthesis, and cigarette smoking. Anthropogenic activities associated with significant production of PAHs leading, in some cases, to localized areas of high contamination include high-temperature (>700°C) pyrolysis of organic materials typical of some processes used in the iron and steel industry, heating and power generation, and petroleum refining. Aquatic environments may receive PAHs from accidental releases of petroleum and its products, from sewage effluents, and from other sources.

Anthropogenic activities associated with significant production of PAHs include coke production in the iron and steel industry; catalytic cracking in the petroleum industry; the manufacture of carbon black, coal tar pitch, and asphalt; heating and power generation; controlled refuse incineration; open burning; and emissions from internal combustion engines used in transportation. PAHs present in the atmosphere enter rain as a result of in-cloud and below-cloud scavenging (Van Noort and Wondergem, 1985). PAHs may reach aquatic environments in domestic and industrial sewage effluents, in surface run-off from land, from deposition of airborne particulates, and especially from spillage of petroleum and petroleum products into water bodies (Prahl *et al.*, 1984). The majority of PAHs entering aquatic environments remain close to the sites of deposition, suggesting that lakes, rivers, estuaries, and coastal marine environments near centres of human populations are the primary repositories of aquatic PAHs.

> *Naphthalene, even in small concentrations is capable of inducing hepatic histopathological alterations which may cause severe physio-metabolic dysfunction leading to death.*

Critical levels of PAHs   Criteria for human health protection and total PAHs, carcinogenic PAHs, and benzo(α)pyrene have been proposed for drinking water and air, and for total PAHs and benzo(α)pyrene in food: in drinking water, 0.01 to <0.2 μg total PAHs/l, <0.002 μg carcinogenic PAHs/l, and <0.0006 μg benzo(α)pyrene/l; in air, <0.01 μg total PAHs/m^3, <0.002 μg carcinogenic PAHs/m^3, and <0.0005 μg benzo(α)pyrene/m^3; in food, 1.6 to <16.0 μg total PAHs daily, and 0.16 to <1.6 μg benzo(α)pyrene daily.

Degradation of PAHs   Among the aromatic fraction, PAHs represent the most recalcitrant and hazardous petroleum hydrocarbons because of their pyrogenic nature and the complexity of assemblages. The most important degradative processes for PAHs in aquatic systems are photooxidation, chemical oxidation, and biological transformation by bacteria and animals (Neff, 1979). The ultimate fate of PAHs that accumulate in sediments is believed to be biotransformation and biodegradation by benthic organisms (EPA 1980). PAHs in aquatic sediments, however, degrade very slowly in the absence of penetrating radiation and oxygen (Suess, 1976), and may persist indefinitely in oxygen-deficit basins or in anoxic sediments (Neff, 1979). PAH degradation in aquatic environments occurs at a slower rate than that in the atmosphere. Animals and microorganisms can metabolize PAHs to products that may ultimately experience complete degradation. The degradation of most PAHs is not completely understood. Those in the soil may be assimilated by plants, degraded by soil microorganisms, or accumulated to relatively high levels in the soil. High PAH concentrations in soil can lead to increased populations of microorganisms capable of degrading the compounds. PAHs are metabolized by of the liver mixed-function oxidases to epoxides, dihydrodiols, phenols, and quinones. The intermediate metabolites have been identified as the mutagenic, carcinogenic, and teratogenic agents (Sims and Overcash, 1983). The activation mechanisms occur by hydroxylation or production of unstable epoxides of PAHs which damage DNA, initiating the carcinogenic process (Jackim and Lake, 1978).

Microbial degradation of PAHs   The recalcitrant nature of many PAHs stems from their low water-solubility, soil-sorption kinetics, constraints to their solubilization and their resistance to normal microbial degradation. Temperature, redox condition, pH, electron acceptors and overall toxicity are the abiotic factors that affect degradation. Several types of PAH-degrading microbes exist including bacteria, fungi and algae (Table 4.2). Bacteria use PAHs as their sole source of carbon and energy, whereas fungi and algae transform PAHs co-metabolically to non-toxic or more readily degraded products.

Table 4.2 Selected genera of PAH-degrading microbes

Microorganism type	Genera
Aerobic bacteria	*Pseudomonas, Vibrio, Alcaligens, Aeromonas, Bacillus, Staphylococcus*
Anaerobic bacteria	*Pseudomonas, Geobacteraceae*
Non-lignolytic fungi	*Cunninghamella, Aspergillus, Candida, Penicillium*
Lignolytic fungi	*Phanerochaete, Bjerkandera, Pleurotus, Coriolus, Dichomitus, Phlebia, Lentinula*
Cyanobacteria	*Oscillatoria, Agmenellum, Anabaena, Nostoc*
Eukaryotic algae	*Selenastrum, Chlorella, Amphora, Navicula, Nitzschia*

1. *Aerobic bacteria in PAH degradation*   Gram-positive and gram-negative aerobic bacteria isolated from soils, natural waters and superficial sediments polluted with PAHs can be used for degradation. Higher molecular weight PAHs such as chrysene and phyrene that contain four benzene rings can be mineralized by bacteria. The initial steps of PAH degradation are catalysed by multicomponent dioxygenases, which introduce two atoms of oxygen to form cis-hydrodiols. Ring cleavage enzymes oxidize PAHs to substrates, which finally enter the tricarboxylic acid cycle. Monooxygenases are also known to oxidize PAHs, however *trans*-dihydrodiol are formed instead.

2. *Anaerobic bacteria in PAH degradation*   Molecular oxygen is not available in many environments since it is rapidly depleted by aerobic microflora. Aerobic degradation is therefore inhibited, because oxygen is not only an electron acceptor, but is also incorporated into the aromatic ring as the first step of degradation. Anaerobic transformations may play a significant role in the removal of PAHs in oxygen-poor environments such as deep soil, ground water or marine and lacustrine sediments. Redox potential is the critical factor in determining the metabolic diversity of microbial oxygen. Three categories of redox conditions lead to different pathways of PAH degradation.

3. *Nitrate-reducing conditions*   Nitrate-reducing condition has been demonstrated in natural samples and pure cultures. PAH removal rates in denitrifying conditions seems to be similar to those of aerobic degradation when cell densities are similar in the presence of excess nitrate and carbon source.

4. *Iron-reducing conditions*   Naphthalene has been found to be degraded in Fe(III)-reducing conditions in some petroleum-contaminated aquifers.

5. *Sulphate-reducing conditions*   It is one of the most important mechanisms of PAH removal from marine sediments, since sulphate reduction serves as the predominant terminal electron-accepting process. Carboxylation of PAHs has been shown as the initial reaction, leading to the formation of benzoate-like intermediates. They are activated by means of coenzyme A ligation and finally mineralized to CO_2 after ring reduction.

6. *Fungi in PAH degradation*   Fungal degradation of PAHs may be important in the detoxification and elimination of PAHs in the environment. The fungus *Cunninghamella elegans*, for example, inhibited the mutagenic activity of benzo(α)pyrene,

3-ethyl cholanthrene, benz(α)anthracene, and 7,12 dimethylbenz-(α)anthracene, as judged by results of the Ames test using *Salmonella typhimurium* (Cerniglia *et al.*, 1985). The rate of decrease in mutagenic activity in bacterial cultures incubated with PAHs was coincident with the rate of increase in fungal metabolism. *C. elegans* metabolized PAHs to dihydrodiols, phenols, quinones, and dihydrodiol epoxides, and to sulphate, glucuronide, and glucoside conjugates of these primary metabolites in a manner similar to that reported for mammalian enzyme systems, suggesting that this organism (and perhaps other fungi) is important in PAH metabolism and inactivation (Cerniglia *et al.*, 1985).

The fungi are able to thrive in high temperatures, providing potential opportunities for high rates of biodegradation.

Non-lignolytic fungi are also able to degrade PAHs ranging from naphthalene to benzo[α]pyrene. White-rot fungi degrade PAHs non-specifically. The fungi are able to thrive in high temperatures, providing potential opportunities for high rates of biodegradation. Lignolytic fungi degrade lignin through the action of extracellular enzymes, i.e., laccase, lignin peroxidases (LiP) and manganese peroxidases (MnP). A fourth group may be the laccase–aryl alcohol oxidase group. Peroxidases use hydrogen peroxide to promote the one-electron oxidation of chemicals to form free radicals. Many chemicals are not directly accessible to the haem group of peroxidases and their direct oxidation cannot proceed further. However certain chemicals can be oxidized to form free radicals by the peroxidases and subsequently oxidize those that cannot be oxidized directly by the enzyme. Non-lignolytic fungi degrade PAHs by the non-specific group involving the cytochrome P-450 monooxygenases. One atom of molecular oxygen is introduced into the PAH molecule by means of cytochrome P-450, yielding an arene oxide that spontaneously isomerizes to form a phenol, which will be further modified to finally form a *trans*-dihydrodiol.

Algae in PAH Degradation

Prokaryotic (cyanobacteria) and eukaryotic (green, red and brown) algae may degrade PAHs in photoautotrophic condition. PAH transformation by algae is seen on shallow aquatic environments such as running waters, lakes or coastal waters. Degradation by cyanobacteria uses monooxygenase-catalysed reactions more similar to those found in eukaryotic cells than that catalysed by true bacteria, whereas some green algae use dioxygenase-catalysed reactions similar to those found in heterotrophic prokaryotes (Castello, 2004).

BIOREMEDIATION OF CONTAMINANTS

5

GARBAGES OF INDIAN CITIES—POTENTIAL ENERGY SOURCE

It is estimated that the production of city garbage in India is about 41,000 tonnes per day, the annual production is about 15 million tonnes. Nonetheless city garbage produced in Mumbai, Chennai and Kolkata is comparable to that of developed countries. In India garbage is dumped in the outskirts of the cities. The Central Mechanical Engineering and Research Institute, Durgapur has established a first pilot plant to produce electricity by using city garbage. The plant has a capacity to use about 50 kg garbage/ha as a result of which about 5KW electricity can be generated. The process of electricity production is the conversion of garbage anaerobically into biogas and in turn into electricity. Nowadays, municipal, agricultural and light industrial wastes are used for conversion into energy by direct burning in refuse fired energy system. The mixture of wood and bark waste burnt directly is collectively called "hog fuel". The hog fuel combination technology has been developed in USA.

Dubey and Maheswari (2005)

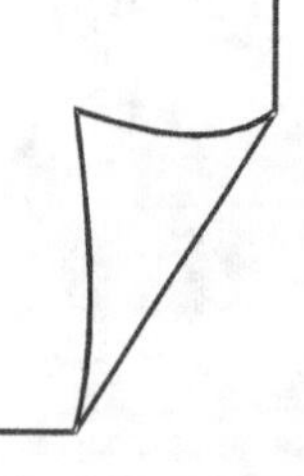

Biodegradable wastes include, organic matter of plant and animal origin, faecal matter, blood, urine, domestic wastes, etc. They are degraded easily and quickly at a faster rate into harmless contaminants. Enormous quantity of sewage and organic wastes pose problems because adequate facilities are not available in cities and towns to handle such large quantities of waste materials. Problems associated with biodegradable contaminants are:

- odorous gases and volatile substances derived from organic wastes

- abundant infectious microbes

- oxygen deficiency caused by organic matter and generation of plenty of plant nutrients in water bodies

ORGANIC WASTES

NATURE OF ORGANIC COMPOUNDS

Organic wastes are abundant in the environment, both as the products of natural and industrial processes. They are generally substituted hydrocarbons that are made up of straight or branched chains, rings or combinations of these structures. Hydrocarbon skeletons are modified by the presence of multiple bonds and substituents (oxygen, nitrogen, halogen, etc.) that markedly influence the reactivity of the molecules. Unsubstituted hydrocarbons are relativity inert, particularly the polar compounds in the environment. These hydrocarbons are thermodynamically unstable with respect to oxidation to CO_2 and H_2O. Organic compounds of soil arising from animal and plant residues are made up of complex carbohydrates, simple sugars, starch, cellulose, hemicelluloses, pectins, gums, mucilage, proteins, fats, oils, waxes, resins, alcohols, aldehydes, ketones, organic acids, lignins, phenols, tannins, hydrocarbons, alkaloids, pigments and other products. Plant residues contain 15–60% cellulose, 10–30% hemicellulose, 5–30% lignin, 2–16% protein and 10% sugars, amino acids and organic acids. Cellulose occurs in a semicrystalline form with a molecular weight of 10^6 and has glucose units with $\beta(1\text{-}4)$ linkages. Hemicelluloses are the polymers of hexose, pentose and sometimes uronic acids with monomers such as xylose and mannose. Lignin is formed by chemical reaction involving phenols and free elements like carbon, hydrogen and oxygen. The molecule is a polymer of aromatic nuclei with either a single repeating unit or several similar units as building blocks.

DECOMPOSITION OF ORGANIC MATTER

The degradation of organic compounds often follows a pathway that is a reverse of synthetic and enzyme-catalysed reactions. Many microbes utilize simple organic compounds for their growth, and their enzymes are responsible for the degradation of these substances. Any compound that is synthesized by a living system is readily degraded biologically and is said to be "biodegradable". However, many of the carbon compounds that have been formed by geochemical reactions (crude oil, coal) or industrial chemical processes (DDT, polyethylene) are not readily degraded.

MINERALIZATION AND IMMOBILIZATION

When microbes grow and multiply on organic debris, carbon is utilized for building the cellular material of microbial cells with the release of carbon dioxide, methane and other volatile substances. During this process, microbes assimilate nitrogen, phosphorus, potassium and sulphur, which get bound in the cell protoplasm. During the decomposition of organic debris, the following processes go on.

1. Degradation of plant and animal remains by cellulase and other microbial enzymes.

2. Increase in the biomass of microbes, which comprises polysaccharides and proteins.

3. Accumulation or liberation of end products.

"**Mineralization**" is the term used to designate the conversion of organic complexes of an element to its inorganic state, which embodies the three processes mentioned above. The other process that involves the microbial uptake of nutrients such as nitrogen, phosphorus and sulphur is known as "**immobilization**".

MICROBES INVOLVED IN DECOMPOSITION

Biodegradable wastes can be effectively treated with microbes and decomposed to simple harmless constituents. Bio-treatment is inexpensive, eliminates pollution problems and provides economically useful products. Though the microbial machinery is unable to handle large amount of organic wastes in the given space, time and conditions, it can be made to work at its maximum efficiency within confined space. This has enabled us to dispose large quantities of biodegradable wastes quickly, efficiently and with very little expenditure. The biodegradable wastes are categorized as liquid and solid. Anaerobic and aerobic processes can degrade the organic wastes. Liquids and semi-solids are subjected to aerobic or anaerobic degradation. Solid wastes are digested and mineralized by anaerobic process. Heterotrophic soil bacteria utilize the available source of organic energy from sugars, starch, cellulose and protein. Autotrophic bacteria use inorganic sources such as iron (*Ferrobacillus*) and sulphur (*Thiobacillus*). Actinomycetes grow on complex substances such as keratin, chitin and other polysaccharides. Heterotrophous soil fungi use organic residues easily. The microbes involved in the decomposition of organic matter are listed in the Table 5.1.

Table 5.1 Microorganisms capable of utilizing different components of organic matter

Nature of substrate in organic matter	Microorganisms
Cellulose	*Alternaria, Aspergillus, Chaetomium, Coprinus, Fomes, Fusarium, Myrothecium, Penicillium, Polyporus, Rhizoctonia, Rhizopus, Trametes, Trichoderma, Trichothecium, Verticillium, Zygorynchus, Achromobacter, Angiococcus, Bacillus, Cellfalcicula, Cellulomonas, Cellvibrio, Clostridium, Cytophaga, Polyangium, Pseudomonas, Sorangium, Sporocytophaga, Vibrio, Micromonopora, Nocardia, Streptomyces* and *Streptosporangium*

(Contd.)

Table 5.1 (Continued)

Nature of substrate in organic matter	Microorganisms
Hemicellulose	*Alternaria, Fusarium, Trichothecium, Aspergillus, Rhizopus, Zygorynchus, Chaetomium, Helminthosporium, Penicillium, Coriolus, Fomes, Polyporus, Bacillus, Achromobacter, Pseudomonas, Cytophaga, Sporocytophaga, Lactobacillus, Vibrio* and *Streptomyces*
Lignin	*Clavaria, Clitocyle, Collybia, Flammula, Hypholoma, Lepiota, Mycena, Pholiota, Arthrobotrys, Cephalosporium, Humicola, Pseudomonas* and *Flavobacterium*
Starch	*Aspergillus, Fomes, Fusarium, Polyporus, Rhizopus, Achromobacter, Bacillus, Chromobacterium, Clostridium, Cytophaga, Micromonospora, Nocardia* and *Streptomyces*
Pectin	*Fusarium, Verticillium, Bacillus, Clostridium* and *Pseudomonas*
Inulin	*Penicillium, Aspergillus, Fusarium, Pseudomonas, Flavobacterium, Beneckea, Micrococcus, Cytophaga* and *Clostridium*
Chitin	*Fusarium, Mucor, Mortierella, Trichoderma, Aspergillus, Gliocladium, Penicillium, Thamnidium, Absidia, Cytophaga, Achromobacter, Bacillus, Beneckea, Chromobacterium, Flavobacterium, Micrococcus, Pseudomonas, Streptomyces, Nocardia* and *Micromonospora*
Proteins and nucleic acids	*Bacillus, Pseudomonas, Clostridium, Serratia* and *Micrococcus*
Cutin	*Penicillium, Rhodotorula, Mortierella, Bacillus* and *Streptomyces*
Tannin	*Aspergillus* and *Penicillium*
Humic acid	*Penicillium* and *Polystitus*
Fulvic acid	*Poria*

ANAEROBIC DECOMPOSITION OF ORGANIC MATTER

Mesophilic and thermophilic microbes decompose organic residues producing carbon dioxide, hydrogen, ethyl alcohol and organic acids such as acetic, formic, lactic, succinic and butyric acids. Mesophilic bacterial flora belonging to the genus *Clostridium* are more active than fungi or actinomycetes. Mesophilic and thermophilic microbes in the compost heaps break down cellulose substrates. After the initial breakdown of complex organic matter, strict anaerobic methanogenic bacteria act upon the secondary substrates (Figure 5.1). Four genera of methane bacteria have been isolated: *Methanobacterium, Methanobacillus, Methanosarcina* and *Methanococcus.* Subba Rao (2000) reported several types of reactions, which can produce CH_4 (Figure 5.1).

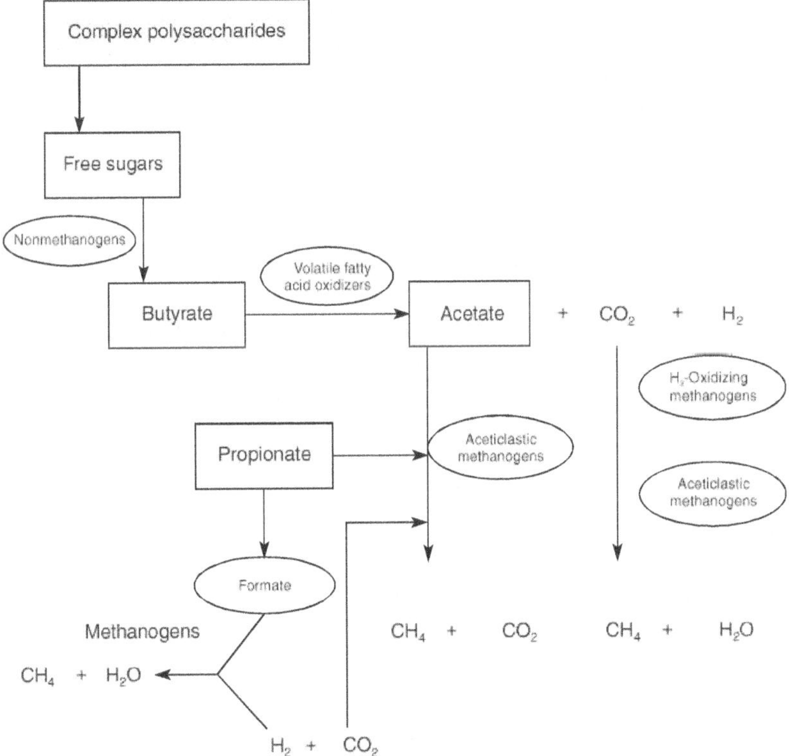

Figure 5.1 Different reactions that can produce CH_4 (Subba Rao, 2000)

> *Mesophilic and thermophilic microbes in the compost heaps break down cellulose substrates.*

HUMUS

The dark coloured and fairly stable soil organic matter formed after the decomposition of organic matter is humus. It can be defined as a ligno-protein complex or an amino acid–lignin complex containing approximately 45% lignin compounds, 35% amino acids, 11% carbohydrates, 4% cellulose, 7% hemicellulose, 3% fats, waxes, resins and 6% other miscellaneous substances including plant growth substances and inhibitors. Raw humus does not support the growth of soil microbes, whereas "nutrient humus" of fertile soils supports microbial growth. Biochemical studies of humus degradation are difficult due to the procedural constraints in solvent extraction of organic matter complexes. Further, biodegradation products of lignin are closely linked with humus substances. Fulvic acid, humin and humic acid have been recognized as the main components of humus complexes. Fulvic acid is the alkali as well as acid soluble part of soil organic matter and contains carbohydrates and proteins. Humic acid is the main component of humus complex and regarded as polymer of aromatic compounds. Fungi, mostly basidiomycetes, are capable of decomposing humic acids. Microbial growth and proliferation can be enhanced by the addition of humic acids. Humus addition increases the growth *of A. niger, P. glaucum, B. mycoides* and *Scendesmus* sp. Humic acid increases the amount of nitrogen fixation by *Azotobacter.* Humus increases the growth of nodulating rhizobial strains such as *R. trifolii, R. meliloti, R. leguminosarum* and *R. japonicum.*

LIGNIN

Lignin is a natural aromatic polymer found in wood and other vascular tissues. It functions as a glue between cells, gives strength to the plant, serves as a barrier against microbial attack and provides an impermeable seal across cell walls of xylem tissue. Lignin originates from the alcoholic precursors ρ-coumaryl and its methoxy substitute counterparts, coniferyl alcohol and synapyl alcohol, which undergo polymerization through free radical attack. The enzymes involved in biodegradation of lignin must be extracellular, non-specific and non-hydrolytic due to its unusual structure. Anaerobic biodegradation of lignin does not exist and the effective degradation is due to a white rot fungi.

> *The enzymes involved in biodegradation of lignin must be extracellular, non-specific and non-hydrolytic due to its unusual structure.*

Microbes for Lignin Degradation

The white rot fungi degrade lignin more rapidly and extensively than other microbes. Generally, the enzymes that play a vital role in biodegradation include ligninases, Mn peroxidases, phenol oxidases and hydrogen peroxide-producing enzymes. However, the achievement of lignin mineralization probably involves a number of electron transfer steps, which also include intracellular enzyme systems. *Pleurotus ostreatus* produces three groups of oxidizing enzymes—laccases, Mn-peroxidases and hydrogen-peroxide— essential for lignin degradation. Other enzymes involved in lignin degradation by *Pleurotus* spp. are acidic and neutral peroxidases, veratyl alcohol oxidases, aryl alcohol oxidases and polyphenol oxidases.

Anaerobic Degradation of Lignin

Lignin (aromatic polymer) constitutes one quarter of lignocellulose, the main component of the earth's renewable biomass resource. Anaerobic degradation occurred when the solubilized lignin fractions were incubated with an inoculum from an anaerobic, mesophilic sludge digester. The highest molecular weight was reduced after 30 days. The reduction was due to the cleavage of inter-monomeric bonds produced during anaerobic attack. Lignin monomers were readily degraded to methane and CO_2 (Kalaichelvan, *et al.,* 1997). Catechol, vanillin, syringic acid, syringaldehyde, ferulic acid, cinnamic acid, *p*-hydroxy benzoic acid, protocatechuric acid and pyrogallol are the products of lignin degradation. Aromatic ring formation of cyclohexanone and cleavage leading to a dicarboxylic acid will degrade the products.

> *Pleurotus ostreatus* produces three groups of oxidizing enzymes—laccases, Mn-peroxidases and hydrogen peroxide—essential for lignin degradation.

CELLULOSE

Cellulose is a β-1, 4-linked polymer that occurs in several forms. It is found in crystalline as well as amorphous forms and occurs in conjugation with other oligosaccharides in the walls of plants and fungi. The ubiquitous distribution of cellulose in municipal, agricultural, and forestry wastes emphasizes the potential use of this material for conversion to useful products. A complex of three basic types of enzymes are essential for cellulose degradation. An *endo*-β-1, 4-glucanase cleaves cellulose to produce smaller oligosaccharides with free chain ends. An *exo*- β-1, 4-glucanase cleaves cellobiose units from the non-reducing chain ends. The cellobiose is then hydrolysed to glucose by a β-glucosidase (Figure 5.2). In most cellulolytic organisms, this complex of enzymes is inducible and is subject to catabolite repression. The cellulose complex remains in association with the cell and functions as a cell-bound, multienzyme complex or *cellosome* (Moat and Foster, 1999).

PECTIN

Pectin is a highly methylated form of poly-1, 4-D galacturonic acid present in plant cell walls. Bacterial phytopathogens such as *Erwinia chrysanthemi* and *E. cartovora* and fungal phytopathogens such *as Monilinia fructigena, Cladosporium cucumerinum* and *Botrytis cinerea* degrade pectin and its derivatives (Figure 5.3). *E. chrysanthemi* and *E. cartovora* are induced to form a complex of enzymes that constitute the degradation of pectin. Initially pectin is demethylated to pectic acid (polygalacturonate) by pectin methylesterase. Chatterjee *et al.* (1988) reported that the major pathway of polygalacturonate catabolism is by bacteria.

Enzymes such as pectate lyase (an endopectate lyase), exopolygalacturonase and oligouronide lyase are involved in the degradation of pectin. Enzymes that degrade pectin or polygalacturonic acid have been found in *Pseudomonas, Bacillus, Clostridium, Lachnospina, Butyrivibrio,* and *Bacteroides.* Members

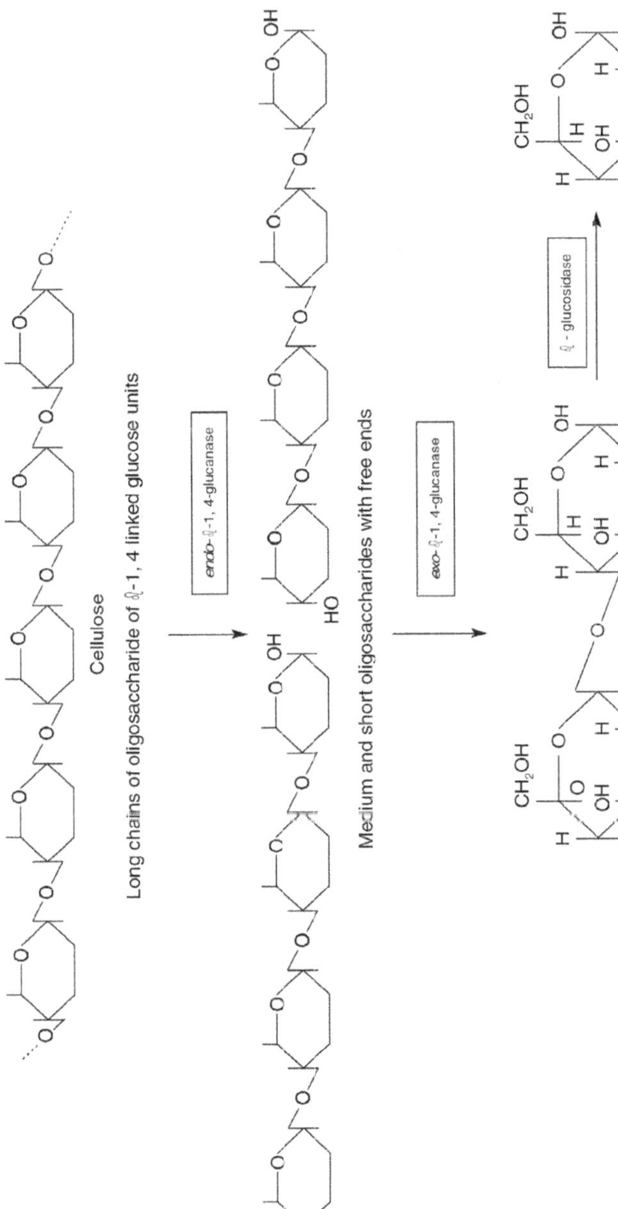

Figure 5.2 Enzymatic degradation of cellulose

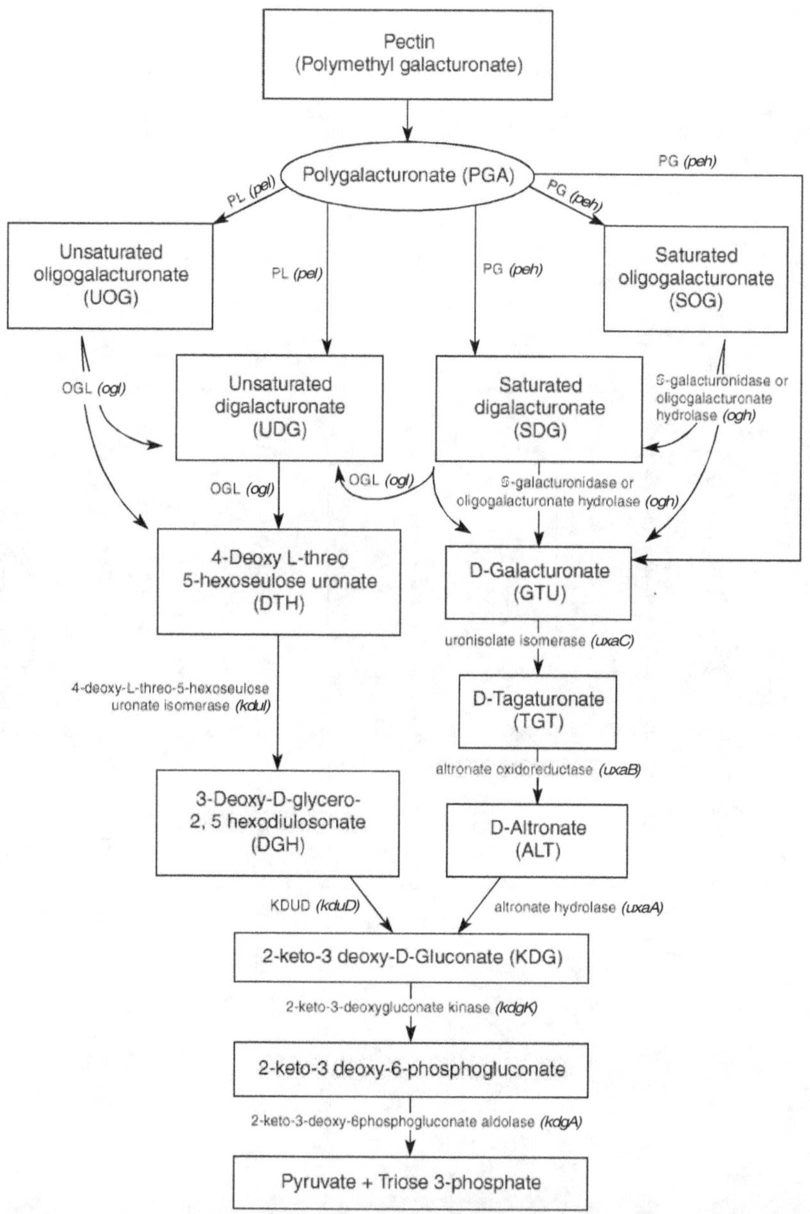

Figure 5.3   Major pathways of polygalacturonate catabolism in bacteria (Chatterjee, Thurn and Tryell, 1985)

> *Pectin is a highly methylated form of poly-1, 4-D galacturonic acid present in plant cell walls.*

of clostridia produce large amounts of butyric acid and appreciable amounts of neutral volatile compounds from pectin. The fungus *Botrytis cinerea* produces soft rot of the parenchyma of carrot root and apple fruit as a result of its ability to produce multiple isozymes of polygalacturonase.

Genetics of Pectin Degradation

Galacturonate utilization pathways and their structural genes are the permease *(exuT)*, uronic isomerase *(uxaC)*, altronic oxidoreductase *(uxaB)*, and altronic hydrolase *(uxaA)*. These genes form the *exu* regulon. The additional structural genes for glucuronate utilization include those for mannonic oxidoreductase *(uxuB)*, and mannonic hydrolyase (*uxuA*). The final stage of the pathway involves 2-keto-3-deoxy D-gluconic kinase *(kdgk)*, and 2-keto-3-deoxy-6-phosphoglucuronide requires the enzyme β-glucuronidase (structural gene, *uidA*) to initiate the pathway (Moat and Foster, 1988).

BIODEGRADABLE LIQUID WASTES

WASTE WATERS

Water which is used for domestic or industrial purposes is considered useless from the point of further use and such disposable water is called as waste water. Under RCRALDR regulation, wastes with <1 wt% total organic carbon (TOC) and < 1 wt% total suspended solids are considered to be waste waters and all other wastes are considered to be non-waste waters. The water emerging out of industry after use is termed as industrial effluent, while water which emerges out after domestic consumption in toilet largely in form of sewage is called as domestic effluent or sewage (Khopkar, 2004). The principal elements present in the organic matter and, which are the main constituents of waste water are carbon, hydrogen, nitrogen, sulphur and phosphorus. The principal substances present are protein, carbohydrates, fats, urea, amines, amino acids, and their products of decomposition. The organic substances present in waste water are unstable and decompose readily through combined chemical and biological actions. The organic matter which bacteria, under biological action, can decompose is called biodegradable and it is of two types: aerobic decomposition or aerobic oxidation and anaerobic decomposition or putrefaction (Deswal and Deswal, 2004).

> *Under RCRALDR regulation, wastes with <1 wt% total organic carbon (TOC) and < 1 wt% total suspended solids are considered to be waste waters and all other wastes are considered to be non-waste waters.*

ENVIRONMENTAL IMPACT OF FERTILIZERS

The major source of nitrogenous wastes can be considered in two groups, chemical and biological. The main chemical sources are fertilizer production, explosives manufacture, coal conversion and the manufacture of sodium carbonate, petroleum refining and refrigeration plants using ammonia. The biological source includes sewage, animal husbandry, food and fermentation industries and natural sources. Nitrogen pollution causes health hazards that have been attributed to nitrate, either directly as a causative factor of methaemoglobinaemia, or indirectly as the source of carcinogenic nitrosamines (Haymar *et al.*, 1991). However the possible damage caused to the ozone layer of the earth's atmosphere by gaseous oxides of nitrogen, released from nitrate ion in the soil is also considered important. Fertilizer efficiency is less than 50% and the fertilizers not used by the crops may have an adverse effect on the environment (Gangea–Cabrera *et al.*, 2003). Nitrates (NO_3^-) lixiviate and pollute water, contributing to eutrophication of lakes and waterways, and making it unacceptable for human consumption. Several gases (NH_3, N_2O and NO), which are produced by volatilization, nitrification and/or denitrification, have been identified as atmospheric contaminants. Some absorb thermic radiation (increasing greenhouse effect); others are partially responsible for the destruction of the stratospheric ozone layer and for acid rain. Nitrogen cycle in agricultural systems is of a very complex nature (Figure 5.4) and depends upon many factors, both biotic and abiotic (Van cleemput and Hera, 1994).

Phosphorus removal from municipal and industrial waste water is required to protect water bodies from eutrophication. Municipal, agricultural and forest wastes are each responsible for about 30 kg of the total known sources of phosphates with detergents contributing about 60% of the municipal phosphate effluent. Biological removal of nitrogen can operate under relatively wide pH variations, it does not create secondary pollution problems and the biomass can be exploited commercially (Proulx and Noire, 1988). The biological process offers the advantage of not requiring addition of chemicals and of reducing the volumes of sludge produced.

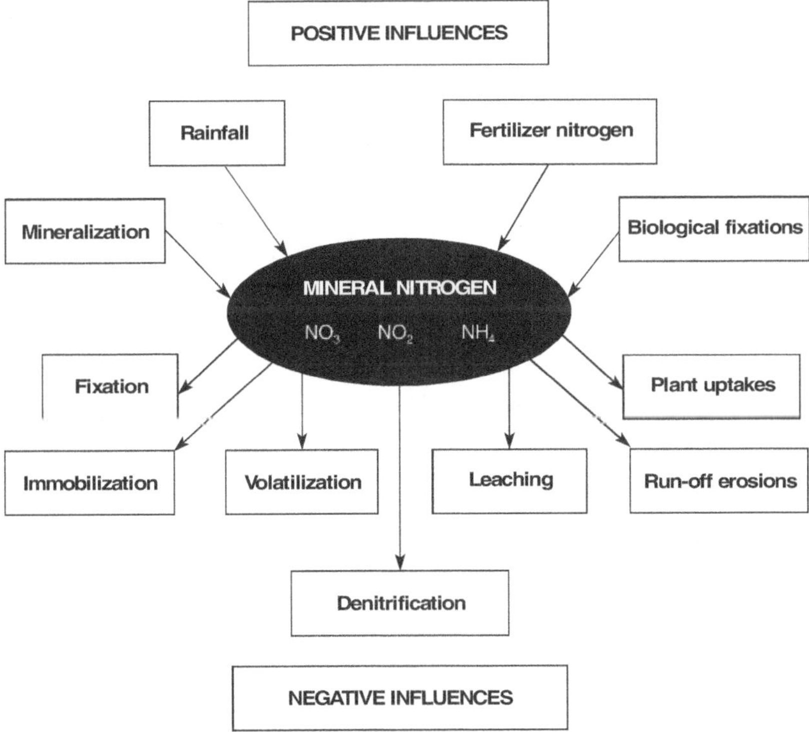

Figure 5.4   Factors influencing the plant-available mineral nitrogen (Van cleemput and Hera, 1994)

> *Phosphorus removal from municipal and industrial waste water is required to protect water bodies from eutrophication.*

TREATMENT AND DISPOSAL OF WASTE WATERS

The present procedures that are available for the removal and management of waste water, involve technical inputs. High operation cost, sludge production, requirement of sophisticated instrumentation and high-energy, and precipitation as hydroxides and sulphides are the constraints of the conventional effluent treatment processes. Current technologies are not only costly and inefficient, but they also increase the exposure of clean-up crews to contaminants. The pollutants are either recalcitrant to biodegradation, or the mineralization rates are very slow due to their toxicity to the organisms. However biological treatment systems have been the preferred treatment option compared to the physical or chemical methods because they are environmentally and economically attractive.

TREATMENT OF DOMESTIC SEWAGE

The main objective of treating sewage of domestic origin is to reduce toxicity and the possibility of spread of diseases. Domestic sewage contains 1% solids and 70% organic matter consisting of fats, carbohydrates and proteins. They are responsible for bad odour and growth of disease-producing microbes. Bacteria, algae and protozoa promote biological degradation, and the oxidation of organic matter provides them food. The biological treatment of sewage depends on factors like susceptibility of organic substances to biochemical oxidation, concentration of the contaminants, pH of biological media, absence of surface active substances and presence of micronutrients like nitrogen, phosphorus, potassium, carbon, etc. The treatment process involves adsorption of contaminants or activated sludge, and mineralization within the microorganisms. Conversion of ammoniacal as well as nitrite nitrogen to nitrate nitrogen is the important change occurring during treatment. Biological treatment of sewage is done in three stages—primary, secondary and tertiary.

> *Biological treatment of sewage is done in three stages—primary, secondary and tertiary.*

Primary Treatment

Before biological treatment of liquid wastes, solid substances and floating objects have to be removed by passing the waste through screens or nets in chambers called screening chambers. After screening, the waste waters are run through a series of grit chambers containing large stones, which can remove the coarse particulate material. Special mechanical devices known as hydrocyclones or centrifugal separators can also be used. The circular motion of the hydrocyclones can separate coarse particles due to the centrifugal force. Clear water is drawn out from the centre of the tank. The remaining fine particulate materials are removed by stocking the waste water in large tanks for variable period of time depending upon the nature of the material. As the lighter particles float and heavier ones settle down, liquid is drawn out from the middle layers of the sedimentation tank. These tanks combine the process of floatation as well as sedimentation. The floatation process can be made more effective by adding emulsifying or foaming agents and agitation with the help of compressed air or mechanical force. The process by which particles get adsorbed onto the surface of the bubbles and come up on the surface is froth floatation. To facilitate the process of sedimentation, finely powdered material known as coagulant can be added (salt of iron, aluminium, activated silica, etc.). The coagulants or flocculants may improve the efficiency of the process but later interfere with the process of biodegradation in the secondary phase of treatment.

Secondary Treatment

Dissolved organic matter, very fine organic debris and particulate materials are removed by secondary treatment which can be categorized as aerobic and anaerobic. The biological process promotes coagulation and oxidation, leading to CO_2 liberation and oxidation of ammonia to nitrites and ultimately nitrates. In aerobic decomposition, oxygen is the final acceptor of hydrogen to form water and more energy is released. In the absence of oxygen, aerobic bacterial cells die for want of energy.

Aerobic decomposition Biodegradable organic matter undergoes aerobic decomposition in the presence of dissolved oxygen. The unstable organic matter in converted into stable and harmless inorganic matter due to the action of aerobic and facultative bacteria. In the first stage carbonaceous matter is oxidized and in the second stage nitrogenous matter is oxidized.

$$\text{Carbonaceous organic matter} + O_2 \xrightarrow{\text{Aerobic bacteria}} CO_2$$

$$\text{Nitrogenous organic matter} + O_2 \xrightarrow{\text{Aerobic bacteria}} NO_3^{2-}$$

$$\text{Sulphurous organic matter} + O_2 \xrightarrow{\text{Aerobic bacteria}} SO_4^{2-}$$

$$\text{Phosphorous organic matter} + O_2 \xrightarrow{\text{Aerobic bacteria}} PO_4^{2-}$$

The end products (carbon dioxide, nitrates, sulphates and phosphates) are stable and inoffensive. Water, heat and additional bacteria will also be produced in aerobic decomposition. The entire process can be represented by the following equation.

i. **Catabolism**

$$\text{Organic matter} + O_2 \xrightarrow{\text{Aerobic bacteria}} CO_2 + H_2O + NH_3 + \text{Energy}$$

ii. **Anabolism**

$$\text{Organic matter} + \text{Catabolic energy} \xrightarrow{\text{Aerobic bacteria}} \text{New bacterial cells}$$

iii. **Autolysis** Self destruction takes place, when the organic matter (nutrients) is exhausted

$$\text{Aerobic bacteria} + O_2 \longrightarrow 5O_2 + NH_3 + 2H_2O + \text{Energy}$$

Aerobic treatment involves biodegradation in the presence of oxygen, which is usually carried out in trickling filters and oxidation ponds.

Dissolved organic matter, very fine organic debris and particulate materials are removed by secondary treatment which can be categorized as aerobic and anaerobic.

Trickling filters   The percolating filters, i.e., sprinklers or tricklers, are useful for biological treatment of sewage at high dosage and loading. The effluent from percolating filter contains "Hums sludge". The hydrostatic head removes sludge without disturbing liquid sewage. Sewage is sprayed over the filter beds and allowed to trickle down through a thick layer of sand and gravel, prepared over a perforated plate. As it moves down, the microbial populations in the different zones of the filter bed cause rapid oxidation or mineralization of organic matter content of the liquid. Trickling filters are equipped with means to force compressed air through pipes under the filter bed that provide oxygen to the microbial population. Bacteria, green and blue-green algae and diatoms grow in the upper zone where plenty of light is available. In the lower zone, fungi, bacteria, protozoans, and small animals, which thrive on dead and decaying organic matter are found. Passage through the biologically active layers oxidizes much organic matter present in the wastes.

Activated sludge process (ASP) involves aeration of settled sewage with special bacterial active sludge leading to clear effluent. This leads to oxidation and nitrification with deposition of brown flocculant sludge. The process is repeated several times to promote purification. Activated sludge purifies sewage in the presence of dissolved oxygen (DO). Golden brown sludge with earthy smell is considered to be good. Sludge activity is measured by rate of oxygen consumption in definite time schedule.

Sludge index is a measure of sludge in terms of volume or density.

$$\text{Sludge Volume Index (SVI)} = \frac{\text{Settled volume of sludge in } \frac{1}{2} \text{ hr}}{\% \text{ of suspended solids}}$$

It increases during settling.

$$\text{Sludge Density Index (SDI)} = \frac{\% \text{ of suspended solids} \times 100}{\text{Settled volume of sludge in } \frac{1}{2} \text{ hr}}$$

The sludge density is obtained as SDI = (2 to 3)

$$\text{Sludge age} = \frac{V \times S_m}{Q \times S_g}$$

where,

V = Tank volume,

Q = Flow of sludge

S_g = Solid in sewage

S_m = Suspended solids in the mixed liquor

It is the average determination of sludge or floc in activated sludge plant. Nitrification can give several benefits such as dense sludge, no bulking, and overload tolerance and easy dewatering. Activated sludge process yields clear, non-putrifiable, odourless effluent with controlled nitrification. Pretreatment such as sedimentation, chemical treatment, oxidation and adsorption is necessary before sludge digestion to reduce load or aeration process. Chlorine gas is injected in sludge to prevent bulking and to promote contraction of humus. Further oxidation is performed in a series of tanks which are known as oxidation ponds.

> *Activated sludge process yields clear, non-putrifiable, odourless effluent with controlled nitrification.*

Oxidation ponds   It involves the slow passage of waste waters, through a large tank in which a large number and variety of microbes are added. Activated sludge is the semi-solid sediment derived from the settlement of particulate matter in waste waters. It comprises a complex mixture of microbes that rapidly decompose most of the organic matter present in water. Compressed air blown through the bottom of the tank provides oxygen necessary for oxidative degradation. The liquid from the oxidation tank is taken out to the settling tank for the settlement of activated sludge and other particulate materials. The clear solution is taken out for disinfection and finally disposed off or subjected to tertiary treatment.

Anaerobic treatment   The anaerobic microbes are capable of decomposing proteins, fats, carbohydrates, cellulose, woody materials, phenols and many other complex substances in the absence or deficit of oxygen. The aim of this process is to change the physical structure of the sludge and to decrease the mass of sludge to promote fermentation of gases and soluble salts and finally to utilize the organic matter as fertilizer. When the free dissolved oxygen is exhausted, aerobic decomposition subsides and anaerobic decomposition begins. Anaerobic and facultative bacteria take part in the process in the absence of dissolved oxygen and light. During anaerobic decomposition, hydrolysis takes place (i.e., hydrogen and oxygen in water molecule are separated and combined with carbon, nitrogen and sulphur. The end products formed are methane, hydrogen, nitrogen, ammonia, carbon dioxide, hydrogen sulphide, organic acids, etc.).

During the first phase of decomposition, large complex organics are broken down into simple organic acids, fatty acids, sugar and glycerol. In the next phase, the bacteria absorb the simple organic substances and new bacterial cells are produced. The end products have offensive odour and are toxic to the organisms themselves. The treatment units that work on anaerobic decomposition or putrefaction are septic tanks, imhoff tanks and sludge digestion tanks.

$$\text{Carbonaceous organic matter} \xrightarrow[\text{bacteria}]{\text{Anaerobic}} CO_2$$

$$\text{Nitrogenous organic matter} \xrightarrow[\text{bacteria}]{\text{Anaerobic}} N_2 + NH_3 + \text{organic acids}$$

$$\text{Organic acids} \xrightarrow[\text{bacteria}]{\text{Anaerobic}} CH_4 + CO_2$$

$$\text{Sulphurous organic matter} \xrightarrow[\text{bacteria}]{\text{Anaerobic}} H_2S$$

Methane formation is one of the processes wherein sludge is separated from sewage and treated. Water acts as an oxidizing agent in methane fermentation. Microbes use hydrogen to reduce CO_2 to CH_4. Oxygen oxidizes the decarbonized radical to form acid. The methane-forming bacteria play an important role in circulation of substances and energy turnover in nature. They absorb CO, CO_2 and H_2 to give hydrocarbon and methane, and synthesize their own cell substances. The solid material left after anaerobic degradation that contains plant nutrients can be used as manure. The septic tank which is a large elongated reservoir with a curved funnel-like bottom, is employed for anaerobic treatment of sewage. The semi-solid sludge left in the septic tank for long duration gets decomposed and its volume is greatly reduced. Periodical stirring and fresh sewage addition will increase the mineralization process. Since it is rich in plant nutrients, it is usually spread out on wasteland. Pathogenic organisms are slowly reduced and the dried scum becomes fit for agricultural purposes. The septic tanks are used in houses to decompose faecal matter under anaerobic conditions. Liquids and semi-solids with a high organic matter content may be used to obtain methane, a fuel gas. Wastes dumped in specially designed compartments produce gases due to decomposition which are trapped over the liquids under the top of the inverted jar. The scum left in the chamber after decomposition of organic matter rich in plant nutrients is dried and finally applied to the fields as fertilizer.

> *Methane formation is one of the processes wherein sludge is separated from sewage and treated.*

Tertiary Treatment

Waste waters containing plenty of nitrates, phosphates and ammonium salts, which the products of biodegradation, may lead to eutrophication or excessive plant growth. The microbial populations in these wastes have to be eliminated before discharging them into rivers or lakes. Biological ponds, a series of simple shallow tanks about 1 metre in depth are used to retain the waste water for the elimination of nutrients and microbes. Algal development absorbs plant nutrients and it supports a series of consumers. The mineral content passes to the higher trophic levels and is eliminated. Algal growth generates oxygen and supports the diverse aerobic communities that consume pathogens. Disinfection and UV treatment makes waste waters fit for discharge in a natural water body.

STRATEGIES FOR BIOLOGICAL TREATMENT OF INDUSTRIAL WASTE WATER

Industrial revolution has generated unprecedented disturbances in the environment due to the introduction of anthropogenic pollutants such as organic, inorganic, and xenobiotic chemicals in the form of untreated industrial waste waters. A majority of the xenobiotics after deliberate or inadvertent release enter the soil and water. Such foreign molecules are either broken down to simpler forms or remain unaltered for a long time because of their persistent nature, posing a great threat to ecosystems, affecting various non-target species and human beings (Mehrotra *et al.*, 2004). It has overwhelmed the self-cleaning capacity of the recipient ecosystems and thus results in the accumulation of pollutants to problematic or even harmful levels. Industrial waste water is more complex than sewage and more difficult to treat since it is heavily laden with organic or mineral matter or with corrosive, poisonous, inflammable or explosive substances

(Sharma, 2003). Industrial waste treatment is thus more challenging, with different effluents having their own problems, and their microbiology and biochemistry being incompletely understood. The engineering aspects of digesters and treatment plants have undergone a lot of progress but their microbiology and biochemistry are mostly neglected areas at least in practice.

Types of Industrial Waste Water

Rapid development of industries has produced a great variety of chemical compounds that have led to the modernization of our life styles. Almost all industries discharge water containing wastes from some stage of their manufacturing process, but industrial wastes are not the same in every case (Figure 5.5). It differs from industry to industry. Industrial wastes are classified as process and chemical wastes. Process wastes are produced due to operations such as washing, operation, washing of raw materials and containers, formation of intermediates and final products, etc. Industrial wastes can be categorized into inorganic (chemical manufacturing units, electroplating industries, metallurgical industries, petroleum industries, etc.) and organic (food processing industries, dairies, breweries, distilleries, paper mills, textile mills and organic chemical manufacturing units) process wastes (Sharma, 2003).

Many industrial units and manufacturing plants, particularly chemical industries as well as industrial processes such as mining, metal plating, metal finishing, rayon manufacture, chrome tanning, etc., to cite a few, discharge sizeable quantities of liquid and solid wastes containing many chemicals and heavy metals into their immediate environment through the plant effluent streams (Iyer, 2003). The importance of industrial waste discharges in relation to water supply and water use is not the same and the problem varies within a specific area, depending upon the distribution and type of industry. The volume of wastes produced varies with the industry and also within each industry. Apart from the variations in the types of industrial wastes and the uneven distribution of the industries, the magnitude of the problem is rarely recognized. The industrial wastes are complex in composition and may contain biodegradable or non-biodegradable or both. Generally biodegradable waste water pollutants are classified as readily degradable organics (e.g. from food industry), complex organics (e.g. from organochemical industry), reactable inorganics (e.g. from heavy industry and plating)

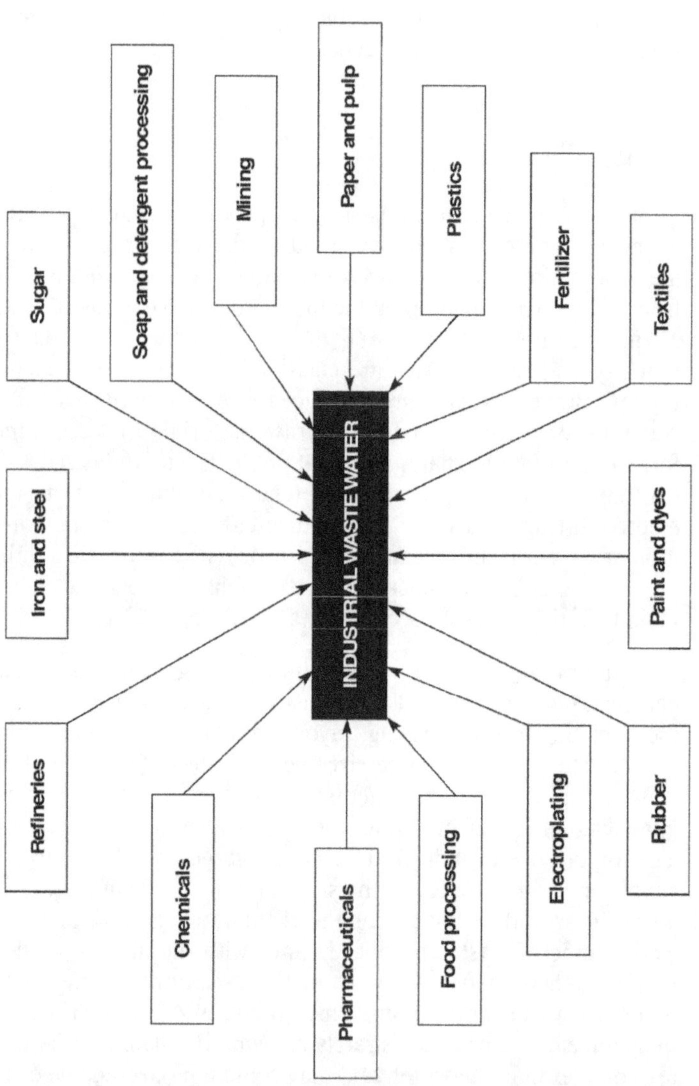

Figure 5.5 Sources of industrial waste water

and inert inorganics (e.g. from coal mining and quarrying). Due to industrial revolution, a wide variety of aromatic hydrocarbons and their derivatives are used in increasing amounts in a number of industrial operations such as the manufacture of chemical solvents, pesticide production and other petroleum-based industries (Jothimani *et al.,* 2003).

Components of Industrial Waste Water

Most of the industries use water in varying proportions. It may be directly used as a solvent and as a medium for chemical reactions or for cooling, washing products and containers, machines, shop floors, etc. The waste water discharged carries with it a number of dissolved and suspended impurities, the composition of which varies with industries and the process used. Industries generally produce wastes containing toxic heavy metals along with organic and inorganic effluents that contaminate the ground water (Table 5.2). The industrial waste waters can be broadly classified as non-fermentable inorganics, fermentable organics and toxic wastes. The effluents discharged by metal industries, steel mills and machine tool factories are non-fermentable wastes and are generally characterized by low pH and high concentration of trace metals. Tanneries, food and meat packing plant effluents contain mostly fermentable wastes. Effluents discharged by electroplating, dyeing, insecticide manufacturing and chemical industries are highly toxic. Many industrial wastes are highly coloured, and are still detectable after high dilution. Wastes from dye manufacturing plants, textile wastes, wood and flax, cooking wastes and fermentation wastes are particularly objectionable from the standpoint of colour. The discharge of toxic materials such as cyanides and chromates affects the self-purification capacity of streams and renders them unfit for drinking purposes. Industrial wastes, in general, do not contain pathogenic bacteria with the possible exception of anthrax organisms found in tannery wastes (Heukelekian, 1997).

The industrial wastes are complex in composition and may contain biodegradable or non-biodegradable components or both.

> *The industrial waste waters can be broadly classified as non-fermentable inorganics, fermentable organics and toxic wastes.*

Table 5.2 Industries generating waste water and their chemical constituents

Waste water generating industry	Constituent of waste water
Acetate rayon	Acetic acid
Atomic energy plant	Fluorides
Breweries	Organic compounds
Chemical manufacturing	Strong acids and bases, spent solvents, reactive wastes
Cleaning agents and cosmetics	Heavy metal dusts, flammable solvents, ignitable wastes, strong acids and bases
Coal	Poisonous gases, carbon particles and solid coal
Construction industries	Ignitable paints, spent solvents, strong acids and bases
Dairies	Sugars
Dyeing	Chromium, nickel and iron
Fertilizer	Low pH, fluorides, phosphate, ammonia and nitrate
Food processing	Acids, organic compounds, suspended solids and oils
Furniture and wood manufacturing	Spent solvents and ignitable wastes
Iron and steel industries	Heavy metals, sulphides, oxides, oil, phenol and naphthol
Launderies	Alkalies, fats, oils, grease and chlorine
Leather products manufacturing	Waste toluene, aromatics, and benzene

(Contd.)

Table 5.2 (Continued)

Waste water generating industry	Constituent of waste water
Metal manufacturing	Heavy metals, cyanides, strong acids and bases
Metallurgical	Iron, acids, mineral acids and metals
Mining	Chlorides, metals, heavy metals, H_2S and H_2SO_4
Nuclear power plants	Radioactive substances
Oil	Oils
Paint manufacturing	Pigments, resins, solvents, lead, aluminium, chromium and zinc
Paper and pulp industry	Cellulose fibres, sulphites, bleaching liquor and organic acids
Pesticide	Chlorides, sulphates and nitrates
Pharmaceuticals	Low pH and chemicals
Photographic processing	High pH, silver, organic and inorganic compounds
Plastics	Phenols, acids and formaldehyde
Plating	Low pH, heavy metals and toxic substances
Printing industry	Heavy metal solutions, waste inks, spent solvents and ink sludge containing heavy metals
Refineries	Oil, phenol and sulphur
Rubber	Variable pH, zinc, cadmium and chimney gases
Soap and detergents	Tertiary ammonia compounds, alkali, fatty acids, glycerol, hydrocarbons and sulphonates
Sugar	Organic matter
Tanneries	Chromium, oil, grease, and sulphides
Textiles	Starch, peroxides, chlorine, hypochlorous acid, alkalies, silicates, phenolic substances and detergents
Vehicle maintenance shops	Spent solvents, reactive wastes, strong acids and bases

Effects of Industrial Waste Water

Industrial waste waters containing chemicals and other harmful products affect the soil, plants and aquatic life through various sources (Figure 5.6). In India, industrial effluents are discharged without proper treatment in the front or rear of the factories. The untreated waste waters move through the unlined channels to the nearest depressions and percolate, resulting in pollution of ground water. In some cases, effluents are discharged into the public sewers, where industrial wastes get mixed with sewage and it is for irrigation. Industrial pollutants accumulate within the biota affecting the fitness of organisms with a high body load of pollutants and its predators and contaminating human food sources. They enter into man's body and cause a number of physiological disorders.

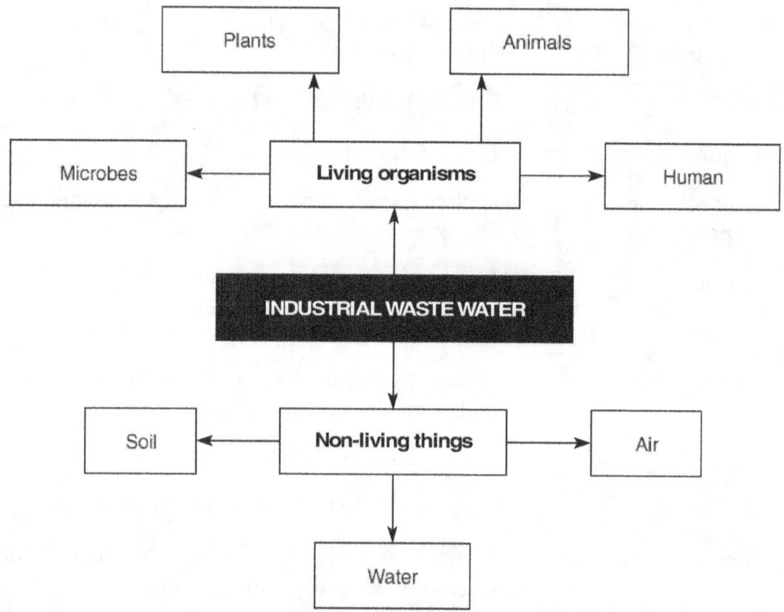

Figure 5.6 Industrial waste waters on living/non-living target components of the biosphere

A direct effect of organic pollutants present in waste and waste water is to deplete the dissolved oxygen (DO) content of the receiving waters due to excessive organic load.

The various living components of the biosphere which are affected by industrial waste waters are listed below.

1. **Microbes**   Inorganic salts, especially nitrogen and phosphorus which are the main constituents of industrial waste waters, induce the growth of microscopic organisms in surface waters. The water and soil environment receive a variety of xenobiotic compounds that are foreign to living systems. Microbes in soil and water act upon these compounds by using them directly as substrates for the energy and biomass production (Singh and Dwivedi, 2004). Industrial waste waters have an impact on microorganisms depending on the nature and composition of the constituents. This can be:

   i.   inhibition of biodegradation

   ii. biodegradation

   iii. incidental removal

   iv. co-metabolism and

   v.   low concentration of substrate situation (Jogdand, 2004).

   The effect of industrial wastes on bacterial population is indirect—it either results in the of multiplication of bacteria deriving their food from the organic matter or in the destruction of bacteria by toxic materials present in the waste.

2. **Plants**   Inorganic salts which are present in most industrial wastes as well as in nature cause water to be 'hard' and make the streams undesirable for industrial, municipal and agricultural usage. Floating solids of waste waters including oils, greases and other materials make the river unsightly and also obstruct passage of light through water, retarding the growth of vital plant foods. The colour of the waste water interferes with the transmission of sunlight into the stream and lessens photosynthetic action. In the USA, an area of at least 30 square miles was contaminated with wastes from manufacture of defoliants, pesticides and chemical warfare agents causing irrigated crops to die.

3. **Animals**   Waste waters discharged from industries change physical, chemical as well as the biological state of water and cause temporary or long-term effects on living organisms. A direct effect of organic pollutants present in waste and waste

water is to deplete the dissolved oxygen (DO) content of the receiving waters due to excessive organic load. The deoxygenation may be high enough to destroy the aquatic life. The water also causes the sporadic outbreak of waterborne diseases because it supports the growth of pathogenic microorganisms (Mehrotra *et al.*, 2004). Suspended solids settle to the bottom or wash on the banks and decompose, causing odours and depleting oxygen in the river water. Solids that settle to the bottom tend to cover spawning grounds and inhibit propagation.

> *The effect of industrial wastes on bacterial population is indirect—it either results in the multiplication of bacteria deriving their food from the organic matter or in the destruction of bacteria by toxic materials present in the waste.*

Direct acting pollutants of industrial waste water create a lethal effect either through (a) their action on the epithelial surface of the gills, or (b) by absorption through the body affecting the internal structure and metabolism of fish. Oil wastes, acids and salts of heavy metals and phenols may produce their lethal effect by acting upon the mucous of the gills by clogging and coagulating them (Heukelekian, 1997). Acids and alkalies discharged by chemical and other industrial plants make streams unsuitable for aquatic life. Fish and other aquatic life are stifled by lack of oxygen, and the oxygen level, combined with other conditions, determines the life or death of fish. Toxic chemicals like insecticides such as toxaphene, dieldrin and dichlorobenzene are very harmful to fish and aquatic life.

4. **Humans**   Wastes generated by industrial activities have brought danger to human beings and very little is known about the long-term consequences of their exposure to the chemicals. Tens of millions of tons of toxic or otherwise hazardous substances enter the environment every year. They cause cancer, delayed nervous damage, malformations in urban children, and mutagenic changes. Once the waste waters containing toxic effluents enter into the environment, they spread in a very complex way and may be converted into other substances, which have a different effect. "Minamata disease" episode due do the discharge of methylmercury industrial waste water in the sea of Japan was an evidence to prove contamination of the fish, which in turn caused neurological disorders to nearly two thousand people—about 400 of them are dead.

Conventional Strategies for Waste-water Treatment

Industrialization leads to the alteration of physical, chemical and biological properties of the environment. With increasing population and industrial expansion, the need for treatment and disposal of waste has grown. Chemical, biological and physical waste-water treatment processes are currently the most commonly used methods of treating aqueous hazardous waste. Appropriate waste-water treatment processes depend on five general interrelated criteria: Economical aspects, eco-toxicological aspects, biological aspects, aesthetic aspects and chemical and engineering aspects. Physico-chemical methods of waste-water treatment are invariably cost-intensive and cannot be employed in all industries especially in developing and underdeveloped countries (Murugesan *et al.,* 2003). The hazards of the pollutants in waste water can be reduced by conventional technologies, which involve removal, alteration, or isolation of the pollutant. These techniques typically consist of excavation followed by incineration or containment. However these technologies in many cases do not destroy the contaminating compounds but rather transfer them from one environment or form to another.

Waste-water treatment plants are usually designated as primary, secondary, or advanced treatment plants, depending on the degree of purification (Figure 5.7). Primary treatment plants utilize physical processes, such as screening and sedimentation, to remove a portion of the pollutants that will settle or float, or that are too large to pass through simple screening devices. The main purpose of secondary treatment is to provide BOD removal beyond what is achievable by simple sedimentation. Early advanced treatment facilities were designed primarily for removing nitrogen and phosphorus (the principal nutrients responsible for eutrophication). However, at present the advanced treatment is designed with the additional goal of removing various toxic substances such as metals.

Physical treatment includes gravity separation, phase-change systems such as steam stripping of volatiles from liquid wastes, and various filtering operations including carbon adsorption (Master, 2004). Chemical treatment transforms waste into less hazardous substances using techniques such as pH neutralization, oxidation, and precipitation. Biological treatment uses microorganisms to degrade organic compounds in the waste stream.

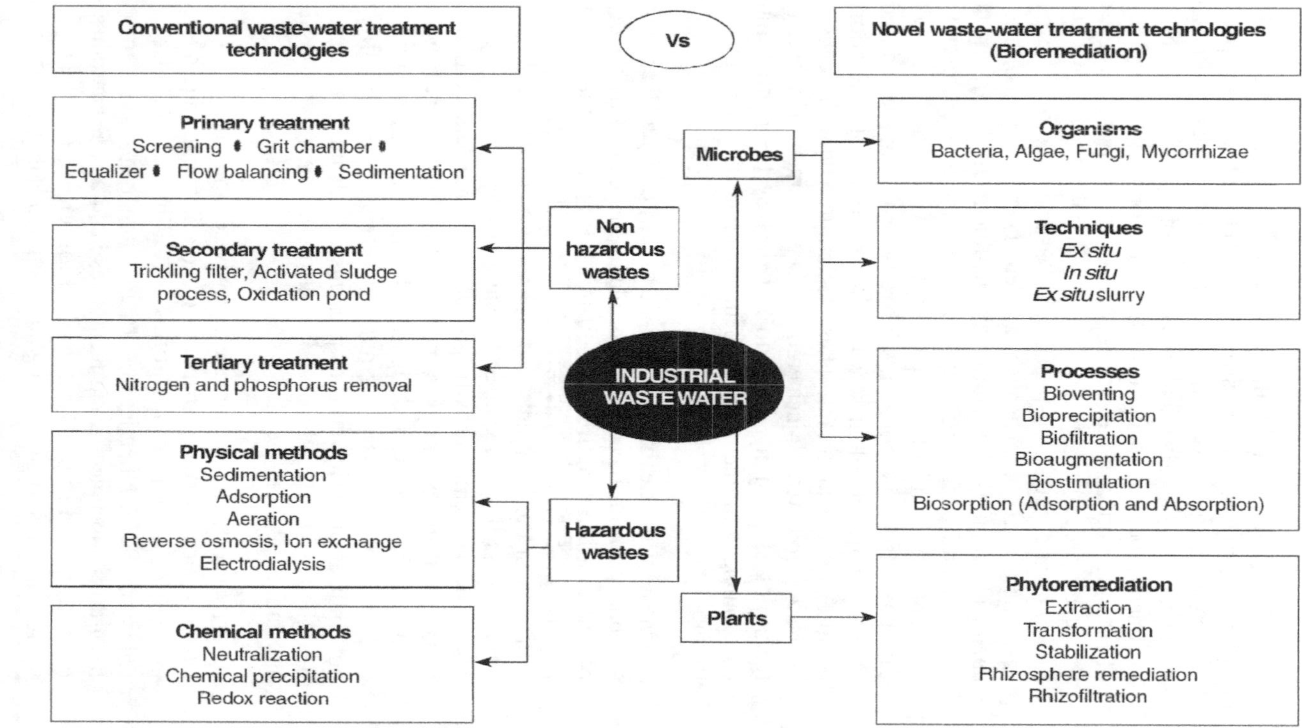

Figure 5.7 Conventional versus novel waste-water treatment technologies

> *Waste-water treatment plants are usually designated as primary, secondary, or advanced treatment plants, depending on the degree of purification.*

Novel Bioremediation Technology for Waste-water Treatment

The main objectives of biological treatment of waste water are to coagulate and remove the non-settleable colloidal solids and to stabilize the organic matter. Appropriate novel bioremediation strategies, using microorganisms and plants can minimize the bioavailability and bio-toxicity of materials found in industrial waste waters. In waste-water treatment, bacteria are primarily responsible for the oxidation of organic matter. However, fungi, algae, protozoa and higher organisms, have important secondary roles to play in the transformation of soluble and colloidal organic matter into biomass which can subsequently be removed from liquid by settlement prior to discharge into natural water source. Industrial waste water can be treated biologically if sufficient nitrogen and phosphorus are present. Biological needs of carbon, nitrogen and phosphorus should be in the ratio 10:5:1.

> *Appropriate novel bioremediation strategies using microorganisms and plants can minimize the bioavailability and bio-toxicity of materials found in industrial waste waters.*

Bioprocess for Waste-water Treatment

Biological degradation is a natural process in which microorganisms make use of pollutants for growth and other cellular processes. Biodegradation processes offer several advantages over physico-chemical options. Microorganisms are self-generating catalysts operating under ambient temperature and pressure conditions. Their reaction specificity permits selective enrichment of microorganisms for a target compound. Microbes have revolutionized the biosorption technology due to their selectivity, short generation time, cost effectiveness, ease in handling and amenability for repeated use. Elevated concentrations of heavy metals discharged into waste water treatment installations may inhibit the sludge biomass from performing its primary function of carbon and nutrient removal (Atkinson *et al.*, 1998). It is therefore of utmost necessity that such nocuous effluents are treated at the source before being discharged into the sewer system.

Bioprocess using microbes such as bacteria, yeasts, algae, actinomycetes and fungi for the synthesis of metal nanoparticles is gaining momentum due to the eco-friendly nature of the organisms which reduce the use of toxic chemicals in conventional strategies. Muralisastry *et al.* (2004) reported that the use of microbes in the biosynthesis of nanomaterials is an amalgamation of curiosity, environmental compulsions, and conviction that has evolved the best process for synthesis of inorganic materials on nano- and macro-length scales.

> *Microorganisms are self-generating catalysts operating under conditions of ambient temperature and pressure.*

Bioreactors for Waste-water Treatment

Bioreactors to treat carbonaceous matter have undergone changes from the original activated sludge and trickling filter systems. A lot of suspended and fixed film versions have come into use. Industrial waste waters, which vary in their volume, load, and complexity, need development of a variety of reactors to achieve efficiency in treatment. The change from aerobic to anaerobic process applications, less sludge-producing, less energy-consuming designs are all reflected in the reactors. With the immobilization of cells and enzymes, bioreactors treating pollutants have actually become smaller in size (Jogdand, 2004).

Microbes for Waste-water Treatment

Conventional waste-water treatment technologies such as incineration and volatilization simply transfer the effluents and create new waste residues. However, in bioremediation technology, detoxification and mineralization of the pollutant to biomass, CO_2 and H_2O make it an attractive, environmentally sound, and cost-effective alternative to conventional procedures. Bioaccumulation refers to uptake and concentration of environmental chemicals from soil, sediments, food and drinking water by diffusion processes. Bioconcentration of xenobiotic compounds readily takes place in bacteria which are exposed to them for a prolonged period because of metabolic adaptations and the production of the enzyme concerned (Singh and Dwivedi, 2004).

Conventional waste-water treatment technologies such as incineration and volatilization simply transfer the effluents and create new waste residues.

Biodegradation of organic compounds of industrial waste waters occurs under anaerobic conditions. Microorganisms are able to use compounds such as nitrate (NO_3^-), sulphate (SO_4^{2-}), or iron (Fe^{3+}) as final electron acceptors instead of oxygen. *In situ* biodegradation is a process wherein bacteria are used to degrade organic compounds in the soil and ground water on site. This relatively new technique has been used most frequently to treat soil contaminated with gasoline and diesel. It shows a great promise in treating chlorinated solvents such as trichloroethylene (TCE), tetrachloroethylene, and 1,2-dichloroethylene (DCE), which are among the most commonly found contaminants in underground water supplies. An obvious advantage to *in situ* biodegradation is that soils and ground water do not have to be removed, so there is less land disturbance, less wasted water, less risk associated with hazardous waste transportation and decreased costs. The approaches to improve *in situ* remediation are as follows.

1. Enhancing the environment of existing microbial populations by adding nutrients and/or oxygen. Nutrients include nitrogen, phosphorus and inorganic salts such as ammonium sulphate, magnesium sulphate, sodium carbonate and calcium chloride. The supply of oxygen can be enhanced by injecting an oxidant such as hydrogen peroxide or by forcing air through wells with diffusers.

2. Altering the underground microbial population by seeding with new microorganisms that have already been acclimated to the pollutants to be degraded (Table 5.3). These microbes can be picked based on laboratory studies of their effectiveness against the waste water in question (Masters, 2004).

Table 5.3 Microorganisms used to treat pollutants in industrial waste water

Industry	Component	Microorganisms
Brewery	Organic chemicals	*Saccharomyces cerevisiae*
Chemical	Halogenated solvents	*Methylotrophous* and *Acetogens*
Dairy	Sugars	*Candida utilis*
Distillery	Spent wash	*Candida utilis*
Dye	Aniline and phenol	*Pseudomonas alcaligenes* and *Micrococcus*
TNT	Trinitrobenzene, TNT	*P. mendocina* and *Citrobacter freundii*
Electroplating	Nickel, chromium	*Bacillus subtilis* and *Aspergillus*
Explosives	Nitroaromatic compounds	*Phanerochaete chrysosporium*
Hospitals	Phenol	*Aureobasidium pullulans*
Mining	Heavy metals	*Azotobacter vinelandii*
Oil	Spilled oil	*Proteobacteria*
Paint	Lead, cadmium	*Acinetobacter anitratus*
Paper	Cellulose	*P. chrysosporium* and *Coriolos versicolor*
Paper	Lignocellulose	*Acetobacterium xylinium*
Pesticide	2,4 -D and dinoseb	*Alcaligenes eutrophus* and *Lipomyces*
Petrochemicals	Hydrocarbons	*Syntrophus* spp. and *Methanosaeta* spp.
Photography	Silver	*A. niger, Fusarium oxysporum*
Pulp mill	Lignin	*Coloriolus versicolor* and *P. chrysosporium*
Radioactive	Uranium	*Rhizopus arrhizus* and *Sargassum*
Sugar	Molasses	*Saccharomyces cerevisiae*
Synthetic fibres	Cyanides	*F. moniliforme* and *Gloeocercospora sorghi*
Tannery	Heavy metals	*Aspergillus* and *Penicillium* spp.
Vegetable processing	Nitrogen	*Nitrosomonas* and *Nitrobacter*

> *In situ biodegradation is a process wherein bacteria are used to degrade organic compounds in the soil and ground water on site.*

Sorption and other Remediation Strategies

Metal contamination can be remediated by approaches involving green plants and their associated microflora, soil amendments and agronomic techniques.

1. **Sorption** The cell wall of bacteria, yeast, algae and fungi which contains several active groups of constituents like acetamido group of chitin, polysaccharide amine (amino peptidoglycoside), sulphahydral and carboxyl groups in protein, phosphodiester (teichoic acid), phosphate and hydroxyl in polysaccharides, participate in biosorption (Gadd, 2000). The use of freely suspended microbial biomass suffers with disadvantages like small particle size, low mechanical strength and difficulty in separating biomass and effluent. The free cells can provide valuable information in laboratory experimentation but are not suited to industrial applications. Immobilized biomass offers many advantages including better reusability, high biomass loading and minimum clogging in continuous flow systems. Biomass immobilized in a range of inert materials like, silica, polyacrylamide, polymethane and polysulphone has been used in a variety of bioreactor configurations, including rotating biological contractors, fixed reactors, trickle filters, fluidized beds and air lift bioreactors (Gadd and White, 1993).

2. **Phytoremediation** Phytoremediation is the use of vegetation for *in situ* treatment of contaminated soils, sediments and water. Plants have shown the capacity to withstand relatively high concentration of organic chemicals without toxic effects, and they can absorb and convert chemicals quickly to less toxic metabolites. In addition, they stimulate the degradation of organic chemicals in the rhizosphere by the release of root exudates and enzymes, and build up organic carbon in the soil. The use of specially selected and engineered metal-accumulating plants for environmental clean-up is an emerging biotechnology, which includes the following

i.   *phytoextraction*   which is the use of metal accumulating plants to remove toxic metals from soil,

ii.   *rhizofiltration*   which is the use of plant roots to remove toxic metals from polluted waters and

iii.   *phytostabilization*   which is the use of plants to limit the bioavailability of toxic metals in soil.

Phytoremediation of heavy metals is designed to concentrate metals in plant tissues from the solid or liquid heavy-metal-laden waste and developing an economical method of extracting the metals from plant residues. This is economical and eco-friendly as it will eliminate the need of costly off-site disposal of the sludge contaminating the ground water.

> *The use of specially selected and engineered metal-accumulating plants for environmental clean-up is an emerging biotechnology.*

3. **Plant–microbe interaction in waste-water treatment**
    Plants can accelerate bioremediation in surface soils by their ability to stimulate soil microorganisms through the release of nutrients from the soil and transport of oxygen to the rhizosphere (Wenzel, 1999). The rhizosphere is rich with a greater number of metabolically active microorganisms than unplanted soil. It is this relationship between soil microbes and the plants that is responsible for the accelerated degradation of soil contaminants.

4. **Genetically engineered organisms in waste-water treatment**   Bioremediation is a treatment process that uses microorganisms to break down or degrade hazardous substances to less toxic or non-toxic substances. This technology is based on the use of naturally occurring or genetically engineered microorganisms (GEMs) to restore the contaminated sites and protect the environment. Packed microbes are used to improve the existing waste-water treatment plants, cleaning oil and solvent spillages, and for degrading toxic and recalcitrant wastes. Use of starter bacterial cultures for waste-water treatment is advantageous, since it reduces the lag phase which is otherwise seen in unadapted indigenous organisms. Microbes from various natural resources are selected for their activity by enrichment

techniques and further genetic manipulation may be done to obtain genetically engineered microorganisms that are efficient degraders of industrial waste-water pollutants (Jogdand, 2004). Samuelson *et al.* (2000) reported recombinant *Staphylococcus xylosus* and *S. carnosus*, which gained Ni^{2+}- and Cd^{2+}-binding capacity, suggesting that these organisms could find use in bioremediation of heavy metals. The nickel-resistant bacteria from anthropogenically polluted biotypes and naturally nickel-rich soils have been hybridized with various probes carrying *cnr* from plasmid pMOL 28 (*A. eutrophus* CH34), *ncc* from pTOM8 (*A. xylosoxidans* 31A) *nre* from pTOM8 (*A. xylosoxidans* 31A) and *nre* from *Klebsiella oxytoca* 15788 (Schmidt and Schlegel, 1994).

Waste-water Treatment by Bioremediation Techniques

Selvaseelan (2001) reported four common *in situ* bioremediation strategies:

1. Pump-and-treat system removal (treatment and return of associated water from a contaminated soil zone).

2. Percolation (applying water containing nutrients and possibly a microbial inoculum to the surface of the contaminated area and allowing it to filter into the soil and mix with the ground water).

3. Bioventing (supply of air to an unsaturated soil zone through the installation of wells connected to associated pumps and blowers, which draw a vacuum on the soil).

4. Air sparging (injection of air into the saturated zone of contaminated soil).

Gadd (2000) suggested that the mechanism of microbial solubilization and immobilization of metal(loid)s, radionuclides and related substances could be successfully integrated with the *in-situ* and *ex-situ* bioremediation processes. Biological approach for metal detoxification offers the potential for selective removal of toxic metals, and operation flexibility and easy adaptability for *in-situ* and *ex-situ* application in a range of bioreactor configurations (Lloyd and Lovley, 2001).

> *Microbial solubilization and immobilization of metal(loid)s, radionuclides* and related substances could be successfully integrated with the *in-situ and ex-situ bioremediation processes.*

Future of Waste-water Treatment Technologies

Bioremediation can become a successful field by integration and proper utilization of natural or modified microbial capabilities with appropriate engineering designs to provide suitable growth environments. Engineered microorganisms, which can be built with special abilities, are providing an important technology to degrade pollutants of industrial waste waters. However a gap exists between advances in laboratory research and commercial applications. Extended screening of microbiota will lead to isolation and identification of microbes with the desired traits. Knowledge of involvement of microbial cellular components in waste-water treatment will help in accelerating the processes of adsorption and absorption. Kinetic data on the interaction of waste water biosorbents are needed for process design, scale-up and evaluating the economics. Effect of environmental factors and nutritional requirements will enable the cultivation of the desirable type of biomass using cheap materials. In the long run, the use of purified biopolymers as specific chemical binding agents holds considerable promise. The application of both genetic and protein engineering techniques could conceivably lead to peptides or other biopolymers with enhanced metal specificity, stability and other useful properties. Identification of cheap and effective immobilizing matrices will help in developing efficient industrial waste water microbial systems for continuous treatment plants. In aquatic systems, plants like *Eichhornia* and *Trapa*, sprouting dense growth of assimilatory roots are expected to afford an efficient means of trapping toxic substances of industrial waste waters.

CONCLUSION

Laboratory level biodegradability studies of industrial waste water is essential to scale up the process at plant level. Problems related to industrial waste-water treatment differs from industry to industry and within the same type of industry. The treatment performance may be influenced by effluent flow rate, frequency of discharge, process difference and the size of the industry. Bioremediation of industrial waste water is an emerging and interdisciplinary technology that involves knowledge of chemical and biochemical engineering, ecology, statistics, microbiology, chemistry and biochemistry and geology, and is considered to be still in development. The forgoing bioremediation research clearly holds promise for effective, economical, and environment-friendly waste-water treatment technologies. The different forms of bioremediation processes such as composting, land farming, bioaugmentation, and waste-water bio-treatment prove to be better in solving the environmental problems of industrial activities. However, the bioremediation strategies for waste-water treatment are still considered to be developing because it is carried out in the natural environment which contains diverse uncharacterized organisms. Another difficulty is that no two environmental problems occur under completely identical conditions. Cheap materials based on natural or waste biomass constitute the basis for new technology that can find its larger application in the detoxification of industrial waste waters. Advancements of bioremediation would benefit from close, interdisciplinary collaborative efforts, from both basic and applied research and may represent a true biotechnology challenge.

SOLID BIODEGRADABLE WASTES

Huge quantities of solid wastes are generated by agricultural, industrial and domestic activities. The amount of solid wastes produced per person per day in Indian cities varies from 300 to 600 g. In India, every year about 410 million tons of manure are formed from organic wastes produced by bovine population (Asthana and Asthana, 2001). Organic wastes are dumped into large compost pits and allowed to stay for a long time. Degradation under anaerobic conditions change the physical structure of the waste material and the mass is reduced. Agriculture and the food processing industry are considered to be the largest contributors to the total annual production of solid wastes.

The amount of solid wastes produced per person per day in Indian cities varies from 300 to 600 g.

CAUSES OF SOLID WASTES

Over-population, affluence, urbanization and technology development are the main reasons for the rapid growth in the quantity of solid wastes (Dhameja, 2004). All solid and semi-solid wastes arising from human and animal activities, except human excreta and sludge are discarded as useless and termed "solid waste" or "refuse". The quantity of solid wastes produced depends upon the living standards of the population and it will be more for an industrialized modern society (Deswal and Deswal, 2004). Rubbish includes combustible and non-combustible solid wastes, excluding food wastes or putrescible materials. Paper, cardboard, textiles, plastics, rubber, wood, etc. are included in combustible wastes. Non-combustible wastes include crockery, tin cans, aluminium cans, metals, construction wastes, etc. Garbage includes putrescible organic wastes like the animal and fruit or vegetable residues resulting from the handling, preparation, cooking and eating as foods.

TYPES AND SOURCE OF SOLID WASTES

The main categories of solid wastes include municipal, industrial and hazardous contaminants (Figure 5.8). The general sources of municipal wastes are the residential, commercial and open areas and include wastes that arise from household activities, restaurants, public places, institutions, markets, street-sweepings, etc. Industrial effluents, rubbish, ashes, construction and demolition wastes mainly constitute industrial wastes that pose a substantial danger immediately or over a period of time to human, plant or animal life. Nuclear plants, hospitals, research institutes and laboratories are the sources of hazardous wastes. Hazardous wastes have characteristics such as ignitability, corrosivity, reactivity or toxicity. Under RCRA and defined in 40CFR 261, hazardous waste is defined as a solid waste that, because of quantity, concentration, or physical, chemical, or infectious characteristics,

- o causes or significantly increases mortality or serious irreversible or incapacitating reversible illness or
- o poses a substantial present or potential hazard to human health or the environment when improperly managed.

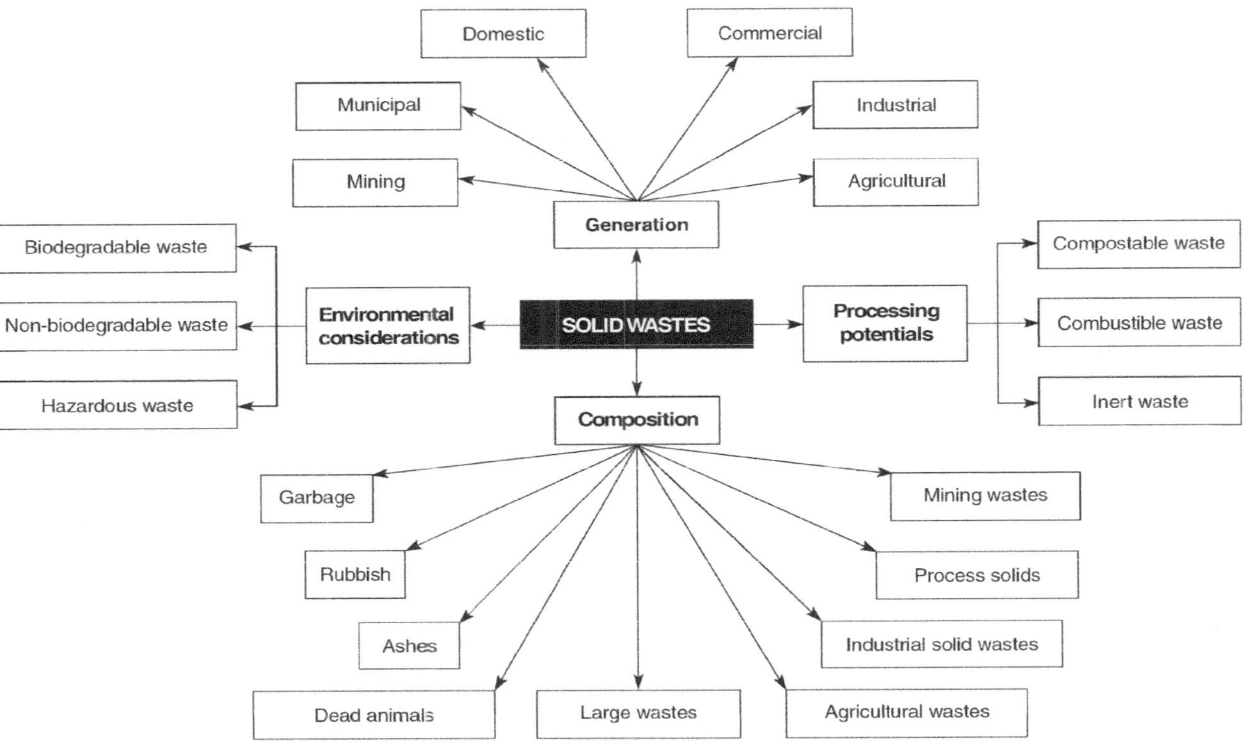

Figure 5.8   Classification of solid wastes

> *Over-population, affluence, urbanization and technology development are the main reasons for the rapid growth in the quantity of solid wastes.*

SOLID WASTE DISPERSAL STRATEGIES

Land filling, incineration, pulverization, composting, pyrolysis and disposal into sea are the current options followed to treat solid wastes. Among these methods, composting is the biological strategy adopted to treat the putrescible organic material in solid wastes. In the process, solid waste is digested anaerobically and converted into human and stable mineral compounds. The volume is reduced considerably and is made free of most of the pathogenic organisms. It is an organic method which converts the solid wastes into manure through anaerobic bacterial action.

Composting and Trenching

Trenches of 4–10 m length, 2–3 m width and 0.7–1.0 m depth are excavated at a clear spacing of 2 m. Solid waste is disposed of into these trenches in layers of 15 cm and by sandwiching 5-cm layers of night soil or animal dung in semi-solid form, till the heaps so formed raise to about 30 cm above the original ground level. A layer of 5–7.5 cm of good earth is then spread on top to prevent access to flies and the wind blowing them off. Within 2–3 days, intensive biological action starts and the organic matter begins to decompose. Heat generated (75°C) during the process prevents the breeding of flies. After about 4–5 months, the decomposed mass gets fully stabilized and changes to brown, odourless innocuous powdery humus having high fertilizer value. It is then removed from the trenches, sieved on 1.25-cm sieves to remove coarse inert materials and sold as manure. The same trench can be used again for receiving further batches of solid wastes.

Open Windrow Composting

The coarse inert matter is removed from solid wastes and then dumped on the ground in the form of piles that are 5–10 m long,

1–2 m wide and 0.5–1.0 m high at about 60% moisture content. The piles are then covered with night soil/animal dung. Due to the biological activity of aerobic bacteria, heat starts developing up to about 75°C in the piles. After a few days, the pile is turned up for cooling and aeration to avoid anaerobic reactions. The process of turning, cooling and aeration is repeated. The complete process takes about 4–6 weeks, after which the compost is ready for use as manure.

Mechanical Composting

In this process, mechanical devices expedite conversion of putrescible organic matter to compost by turning. Mechanical composting stabilizes the solid wastes within 3–6 days. Reception of solid wastes, segregation, shredding or pulverization, stabilization and preparation of the stabilized mass for the market are the various steps involved in large-scale mechanical composting.

BIOREMEDIATION OF SPECIFIC POLLUTANTS

6

BIOREMEDIATION IN ACTION

Biotechnological approaches involving biodegradation is considered more attractive for bioremediation. Companies like Bio Technical are working on treating polluted sites *in situ*. Dr. Ananda Chakrabarty, an Indian-born American scientist, was awarded with patent for *Pseudomonas aeruginosa*, a bacterium designed to biodegrade spilled oil. The bacterium (genetically modified) produced a glycolipid emulsifier, that reduced the surface tension of the oil–water interface and thus helped in removal of oil from water. The microbial emulsifier was non-toxic and biodegradable, thus being ecologically sound. The use of genetically engineered microbes for cleaning spilled oils will also be facilitated by mixing the microbe with straw, which can be stored. When needed, the straw with mixed microbes can be scattered over the spilled oil, so that the straw will first soak oily water and then the microbe will break up the oil into harmless, non-polluting materials, rendering the water and the soil harmless.

Gupta (2004)

SUPER BUG

Super bug is a bacterial strain of *Pseudomonas* that can degrade camphor, octane, xylene and naphthalene. The bacterium containing CAM (Camphor-degrading) plasmid conjugated another bacterium with OCT (Octane-degrading) plasmid. Due to the presence of homologous regions of DNA, recombination occurs between these two plasmids resulting in a single CAM–OCT plasmid. This new bacterium possesses the degradative genes for both camphor and octane. Another bacterium with XYL (Xylene degrading) plasmid is conjugated with NAH (naphthalene-degrading)-plasmid-containing bacterium. XYL and NAH plasmids are compatible and can coexist in the same bacterium. This newly produced bacterium contains genes for the degradation of xylene and naphthalene. CAM–OCT-plasmid-containing bacterium can be conjugated with NAH-plasmid-containing bacterium to produce the superbug which contains CAM–OCT plasmid.

Satyanarayana (2005)

PETROLEUM HYDROCARBONS

NATURAL OIL SEEPS

Natural oil seeps developed in submarine rock strata have been discharging crude petroleum into the environment since millions of years. More than 200 submarine oil seeps have been identified around the world and the amount of crude petroleum discharged annually by them is far greater than the total contamination from offshore oil production and transportation activities. The slow discharge of petroleum crude that occurs inside the ocean is taken care of by natural agencies of degradation. Man-made oil spills, however tend to be disastrous because they introduce suddenly enormous quantities of crude petroleum on ocean surface which causes problems of toxicity to organisms exposed to it and an ecological catastrophe (Figure 6.1). Oil spills are generally associated with the release of hydrocarbons to the oceans or other water bodies, where the oil floats and spreads over large areas of water surface. Slow degradation of petroleum residues affects the aquatic life. The crude oil input into the sea is 7×10^8 gal/year worldwide.

Big spills	37	
Routine maintenance	137	
Down the drain	363	
Up in smoke	92	
Offshore drilling	15	
Natural seeps	62	

Figure 6.1 Petroleum hydrocarbons in marine environments (annually in millions of gallons). **Source** http://seawifs.gsfc.nasa.gov/ OCEAN-PLANET/HTML/peril-Oil_pollution/html.

Huge quantity of petrol has to be transported across the sea and land from one place to another due to its limited distribution to certain places and rock formation. During transportation, some of this material spills out and causes enormous damage to the environment (Table 6.1). Each litre of spilled oil may spread to cover an area of about 4000 sq. metres over the water surface while its oxidation requires about 3.3 kg of oxygen, which usually occurs in four hundred thousand litres of water (Asthana and Asthana, 2001).

Table 6.1   Major oil spills in the order of amount spilled

Date	Source and place	Amount of oil spills (Millions of gallons)
26 January 1991	Terminals, tankers; 8 sources total; sea island installations; Kuwait; off coast in Persian Gulf and in Saudi Arabia during the Persian Gulf War	240.0
03 June 1979	Ixtoc I well; Mexico; Gulf of Mexico, Bahia del Campeche	140.0
02 March 1990	Uzbekistan; Fergana Valley	88.0
04 February 1983	Platform no. 3 well; fran; Persian Gulf, Nowruz Field	80.0
06 August 1983	Tanker *Castillo de Bellver;* South Africa; Atlantic Ocean, 110 km northwest of Cape Town	78.5
16 March 1978	Tanker *Amoco Cadiz*; France; Atlantic Ocean, off Portsall, Brittany	68.7
10 November 1988	Tanker *Odyssey;* Canada; North Atlantic Ocean, 1175 km northeast of St.Johns, Newfoundland	43.1
19 July 1979	Tanker *Atlantic Empress*; Trinidad and Tobago; Caribbean Sea, 32km northeast of Trinidad–Tobago	42.7
11 April, 1991	Tanker *Haven;* Genoa, Italy	42.0
01 August 1980	Production well D-103 (concession well); 800 km southeast of Tripoli, Libya	42.0
02 August 1979	Tanker *Atlantic Empress*; 450 km east of Barbados	41.5

(Contd.)

Table 6.1   (Continued)

Date	Source and place	Amount of oil spills (Millions of gallons)
18 March 1967	Tanker *Torrey Canyon*; United Kingdom; Land's End	38.2
19 December 1972	Tanker *Sea Star*; Oman; Gulf of Oman	37.9
23 February 1980	Tanker *Irene's Serenade*; Greece; Mediterranean Sea, Pilos	36.6
07 December 1971	Tanker *Texaco Denmark*; Belgium; North Sea	31.5
23 February 1977	Tanker *Hawaiian patriot*; United States; Pacific Ocean 593 km west of Kauai Island, Hawaii	31.2
20 August 1981	Storage tanks; Kuwait; Shuaybah	31.2
25 October 1994	Pipeline: Russia; Usinsk (in area that was closed to foreigners before collapse of Soviet Union)	30.7
15 November 1979	Tanker *Independentza*; Turkey; Bosporus Strait near Istanbul, 0.8 km from Hydarpasa port	28.9
11 February 1969	Tanker *Julius Schindler*; Portugal; Potna Delgada, Azores Islands	28.4
12 May 1976	Tanker *Urquiola*; Spain; La Coruna Harbor	28.1
25 May 1978	Pipeline no. 126 well and pipeline; Iran; Ahvazin	28.0
05 January 1993	Tanker *Braer*; United Kingdom; Garth Ness, Shetland Islands	25.0
29 January 1975	Tanker *Jakob Maersk*. Portugal; Porto de Leisoes, Oporto	24.3
06 July 1979	Storage tank tank no. 6; Nigeria; Forcados	23.9

Source   http://www.cutter.com/oilspill

Problems Associated with Biodegradation of Crude Oil

Susceptibility of oil to biodegradation varies due to the complexity of oil. The rate of biodegradation is 100–960 mg of oil/m^3 of sea water/day. Oil biodegradation in marine environment is limited due to: low water temperature, scarcity of hydrocarbon-degrading organisms, scarcity of nutrients, minerals, nitrogen, phosphorus and the toxic compounds of the oil. Oil spills in the sea water are low in nutrients (carbon and nitrogen) and minerals (phosphorus and iron). Oil can serve as carbon source if other elements are available for growth. Spray irrigated special fertilizers that dissolve in oils or inorganic compounds carried in oil-soluble base can be used to remediate oil pollution in marine environment. The main constraints include:

- ○ the native degradative organisms
- ○ foreign organisms must survive in the new environment
- ○ the inoculum must remain in contact with oil

More than 200 submarine oils seeps have been identified around the world and the amount of crude petroleum discharged annually by them is far greater than the total contamination from offshore oil production and transportation activities.

Light and Medium Fraction

They are the mixture of various hydrocarbons containing one to sixteen carbon atoms joined together in straight chains or simple compounds containing an aromatic ring. The lighter of these are gases and volatile substances (vapours) that are poisonous. The toxicity of hydrocarbons are indirectly proportional to their boiling point and viscosity. As the viscosity and boiling point rise, the toxicity decreases. The light and medium fraction of petroleum evaporates quickly due to low boiling point and high volatile nature.

Heavier Fraction

Heavy oils, greases, waxes and solid fraction consist of 10–40% carbon joined together in various ways forming complicated chains, rings, polycyclic and heterocyclic structures and fused ring systems. The structural complexity makes these compounds resistant to natural decay and degradation. The recalcitrant and refractory nature of these compounds makes them important environmental contaminants though they are less toxic. Heavy oils and lubricants spilled or discarded on the soil take months to oxidize and be decomposed. Most of these chemicals are lipophilic in nature and enter into the biological systems causing toxicity.

> *The toxicity of hydrocarbons are indirectly proportional to their boiling point and viscosity.*

Microbial Treatment of Crude Oil

Microbial degradation of oil has been shown to occur by the attack of the aliphatic or light aromatic fractions of the oil. Several microbial species live on hydrocarbons and are responsible for biodegradation of crude petroleum. Natural microbes are capable of decomposing limited number of hydrocarbons as raw materials. Inoculation of hydrocarbon-degrading organisms and addition of nitrogen and phosphorous fertilizers to support their growth is being explored for attack on oil slicks. Oil needs to be brought into soluble physical state for microbial attack. Tarry lumps and stable emulsions (referred to as chocolate mouse effect) are difficult for attack. Fungi and bacteria are the main agents which decompose oil and oil products. Besides cyanobacteria, yeast and algae have shown to oxidize hydrocarbons. Oil is insoluble in water and is less dense; it floats on the surface and forms slicks or oil films. Hydrocarbon-oxidizing microbes develop rapidly in such films. Many pseudomonads, different cyanobacteria and *Mycobacteria* are able to degrade petroleum products. Initially, bacteria oxidize the non-volatile compounds and in the latter process, certain fractions of branched-chain and polycyclic hydrocarbons are degraded slowly (Dubey and Maheswari, 2005).

Pseudomonas decomposes a number of esoteric compounds, which most microbes are incapable of degrading. The genes coding for enzymes which attack hydrocarbons are located at plasmids. Microbial products serve as an efficient replacement for chemical dispersants and surfactants. A strain of *Pseudomonas aeruginosa* produces large quantities of a surface-active agent capable of reducing viscosity and surface tension at the oil–water interface. This being a product of biological activity, is biodegradable. Addition of important nutrients such as phosphorus and nitrogen to oil spills can increase bioremediation rates significantly.

Crude Petroleum and its Distillates

Crude oil is a complex mixture of hydrocarbons, basically composed of aliphatic, aromatic and asphaltene fractions along with nitrogen, sulphur and oxygen-containing compounds. They occur deep inside the earth's crust and are termed as crude petroleum or mineral oil. Though the composition varies according to locality, all of them contain hydrocarbon (aromatic, aliphatic, cycloparaffins, naphthenes, etc.). Crude petroleum also contains metallic constituents like nickel, chromium, cadmium, vanadium, iron, etc. Crude petroleum is used to separate components like petrol and benzene, medium oils such as kerosene and diesel, heavy oils such as lubricating oils, grease, vaseline, hard wax, coal tar, etc.

> *Heavy oils and lubricants spilled or discarded on the soil take months to oxidize and be decomposed.*

Petroleum Degradation

Oil floating on water is destructive to birds and various forms of marine life, and when driven ashore, causes heavy economic and aesthetic damage. Dissolved aromatic components of petroleum disrupt, even at a low ppb concentration, the chemoreception of some marine organisms (Atlas and Bartha, 2005). Petroleum is a complex mixture of aliphatic, alicyclic (cycloaliphatic saturated ring structures), and aromatic hydrocarbons and a smaller proportion of non-hydrocarbon compounds such as naphthenic acids, phenols, thiols, heterocyclic nitrogen and sulphur compounds, as well as metallophyrins. Petroleum is a natural product, resulting from the anaerobic conversion of biomass under high temperature and pressure. National Research Council (1985) reports that the production, transportation, refining, and ultimate disposal of petrochemicals are increasing (annually 3.2 million metric tonnes of oil are introduced into the oceans).

> *Natural microbes are capable of decomposing limited number of hydrocarbons as raw materials.*

Pathways of Petroleum Degradation

There are two techniques for utilizing bacteria to degrade petroleum in the soil. One method uses the bacteria that are already found in the soil. These bacteria are stimulated to grow by introducing nutrients into the soil and thereby enhancing the biodegradation process (biostimulation). The other method (bioaugmentation) involves culturing the bacteria independently and adding them to the site (Mathewson, 1988).

The initial attack on alkanes occurs by enzymes that have a strict requirement for molecular oxygen, that is monooxygenases (mixed function oxygenases) or dioxygenases (Singer Finnerty, 1984). In the first equation (1) one atom of O_2 is incorporated into the alkane, yielding a primary alcohol. The other is reduced to H_2O with the reduced form of nicotinamide dinucleotide phosphate ($NADPH_2$) serving as electron donor.

$$R-CH_2-CH_3+O_2+NADPH_2 \longrightarrow$$
$$R-CH_2-CH_2-OH+NADP+H_2O \qquad (1)$$

In the second equation, both atoms of O_2 are transferred to the alkane, yielding a labile hydroperoxide intermediate that is subsequently reduced by $NADPH_2$ to alcohol and water (3).

$$R-CH_2-CH_3+O_2 \longrightarrow R-CH_2-CH_2-OOH \qquad (2)$$

$$R-CH_2-COOH-OOH+NADPH_2 \longrightarrow$$
$$R-CH_2-CH_2-OH+NADP+H_2O \qquad (3)$$

The initial attack is directed at the terminal methyl group, forming a primary alcohol that in turn is oxidized to an aldehyde and fatty acid. Occasionally, both terminal methyl groups are oxidized in this manner, resulting in the formation of a dicarboxylic acid. This variation is described as di-terminal or co-oxidation. It is one of the ways to bypass a block to β-oxidation due to branching of the carbon chain.

Microbial products serve as an efficient replacement for chemical dispersants and surfactants.

If fatty acid is formed, further catabolism occurs by the β-oxidation sequences. The fatty acid is converted to acyl coenzyme-A (4) and a series of enzymes act on it with the result that an acetyl CoA group is cleaved off and the fatty acid is shortened by the two-carbon unit (equations 4–9). This sequence is repeated. The acetyl CoA are converted to CO_2 through the tricarboxylic acid cycle. The end products of hydrocarbon mineralization are CO_2 and H_2O.

$$R-CH_2-CH_2-CH_2-COOH \xrightarrow[+CoA]{} \tag{4}$$

$$R-CH_2-CH_2-CH_2-\overset{\displaystyle O}{\overset{\|}{C}}-CoA \xrightarrow[-2H^+]{} \tag{5}$$

$$R-CH_2-CH=CH-\overset{\displaystyle O}{\overset{\|}{C}}-CoA \xrightarrow[+H_2O]{} \tag{6}$$

$$R-CH_2-\underset{\displaystyle OH}{\overset{\displaystyle }{CH}}-CH_2-\overset{\displaystyle O}{\overset{\|}{C}}-CoA \xrightarrow[-2H^+]{} \tag{7}$$

$$R-CH_2-\underset{\displaystyle \underset{\displaystyle O}{\|}}{C}-CH_2-\overset{\displaystyle O}{\overset{\|}{C}}-CoA \xrightarrow[+CoA]{} \tag{8}$$

$$R-CH_2-\underset{\displaystyle \underset{\displaystyle O}{\|}}{C}-CoA+CH_3-\overset{\displaystyle O}{\overset{\|}{C}}-CoA \tag{9}$$

Anaerobic Petroleum Degradation

The microbial diversity in a hydrocarbon and chlorinated-solvent-contaminated aquifer undergoing intrinsic bioremediation was assessed by cloning and sequencing bacterial and archaeal 16SrDNA fragments. This study detected phylotypes that were closely related to *Syntrophus* spp. and *Methanosaeta* spp. (Ahmad and Zarkaria, 2004). Potential anoxic biodegradation mechanism involves a direct dehydrogenation of an intact hydrocarbon (10), leading through alkane (11), alcohol (12) and aldehyde (13) to fatty acid (14).

$$R-CH_2-CH_3 \xrightarrow[-2H^+]{} \tag{10}$$

$$R-CH=CH_2 \xrightarrow[+H_2O]{} \tag{11}$$

$$\underset{OH}{R-CH=CH_3} \xrightarrow[-2H^+]{} \tag{12}$$

$$R-CH_2-CHO \xrightarrow[-2H^+,\ +H_2O]{} \tag{13}$$

$$R-CH_2-COOH \tag{14}$$

These sequences would allow the anaerobic metabolism of an intact hydrocarbon in the presence of an electron acceptor such as nitrate. Some microbes attack alkanes subterminally; that is oxygen is inserted on a carbon atom within the chain instead of at its ends. A secondary alcohol is formed first, which is further oxidized to ketone and finally to an ester. The ester is cleaved, yielding a primary alcohol and a fatty acid. The alcohol fragment is oxidized through the aldehyde to the fatty acid analog and both fragments are metabolized further by β-oxidation sequence (Singer and Finnerty, 1984).

> *Oil floating on water is destructive to birds and various forms of marine life, and when driven ashore, causes heavy economic and aesthetic damage.*

DEGRADATION OF ACID MINE DRAINAGE

Coal and various metal ores are enclosed in geological formations of a reduced nature. When mining activities expose pyrite (FeS_2) to atmospheric oxygen, a combination of auto-oxidation and microbial iron and sulphur oxidation produces large amounts of acids. The iron-rich acidic mine drainage kills aquatic life and renders the streams contaminated. Atmospheric oxygen oxidizes pyrite at a neutral pH. In the pH range of 3.5 to 4.5, the stalked iron bacterium *Metallogenium* catalyses the reaction. Below 3.5 pH the acidophilic bacterial genus *Thiobacillus* takes care of the reaction. Microbial oxidation at this stage is several hundred times higher than auto-oxidation. The overall reaction for the oxidation of pyrite by *Thiobacillus thiooxidans* and auto-oxidation can be summarized as follows:

$$2FeS_2 + 7\tfrac{1}{2}O_2 + 7H_2O \longrightarrow 2Fe(OH)_3 + 4H_2SO_4$$

The sulphuric acid produced, accounts for the high acidity and brown colour of the effluent. The feasibility of treating mine effluent using the activity of *Desulfovibrio* and *Desulfotomaculum* was demonstrated on a laboratory scale. Mine effluent combined with large amount of organic waste, such as sawdust can increase the activity of aerobic and facultatively anaerobic cellulolytic microbes. They lower the redox potential and produce degradation intermediates that can be utilized by sulphate reducers. The hydrogen sulphide generated, first reduces the ferric iron to ferrous and precipitates the latter as FeS. The process restores the neutral pH and removes the iron and sulphur from the effluent.

> *Mine effluent combined with large amount of organic waste, such as sawdust can increase the activity of aerobic and facultatively anaerobic cellulolytic microbes.*

BTEX

The BTEX chemicals, benzene, toluene, ethylbenzene and *o*-, *m*- and *p*-xylene are the major contaminants of soil and ground water, originating mainly from leaking underground gasoline tanks. Denitrifying (Strain T1) and iron-reducing (Strain GS-15) strains have been isolated that can grow on toluene and *o*-xylene and can transform *m*-xylene, 7,8. The proposed pathway is through oxidation of the methyl substituent to benzoyl coenzyme-A through the involvement of acetyl coenzyme-A, with concomitant formation of benzylfumaric and benzylsuccinic acids as dead-end metabolites involving succinyl coenzyme-A (Figure 6.2). In mixed methanogenic cultures using ^{14}C-labelled toluene, labelled benzene, phenol and *o*- and *p*-cresol as well as methyl cyclohexane and benzoic acid have been observed.

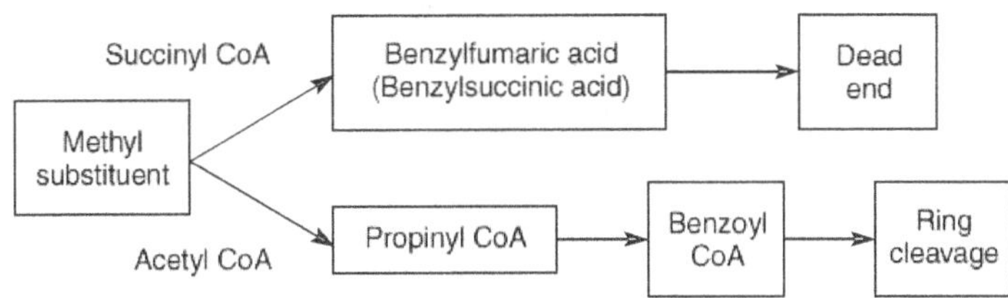

Figure 6.2   Oxidation of methyl substituent to benzoyl CoA

Degradation of Toluene

Pseudomonas putida mt-2 oxidizes the methyl group of toluene to form benzyl alcohol followed by reactions leading to the formation of catechol, the substrate for the metacleavage pathway leading to the formation of an α-keto acid. *Pseudomonas mendocina* oxidizes toluene to *p*-cresol, which undergoes orthocleavage with the formation of ketoadipate (Figure 6.3). *P. putida* PpF1 initiates the oxidation of toluene by incorporating oxygen into the aromatic nucleus to form *cis*-toluene dihydrodiol [*cis*-1(s), 2(R)-dihydroxy-3 methyl cyclohexa 3,5-diiene], which is converted to 3-methylcatechol. 3-methylcatechol is then degraded via the metacleavage pathway to yield an α-keto acid intermediate formed via the metacleavage route (2-keto-4-hydroxyvalerate) is cleaved by an aldolase to form phyrate and acetaldehyde (Moat and Foster, 1988).

> *The BTEX chemicals, benzene, toluene, ethylbenzene and o-, m- and p-xylene are the major contaminants of soil and ground water.*

Genetics of Toluene Degradation

The enzymes for the complete degradation of toluene in *Pseudomonas putida* mt-2 are encoded by a set of genes borne on a TOL (for toluene) plasmid pWWO. Toluene, *m*- and *p*-xylene, 3-ethyl toluene, and 1, 2, 4-trimethyl benzene are oxidized to the corresponding carboxylic acids (benzoate, *m*- and *p*-toluate, 3-ethylbenzoate, and 3, 4-dimethylbenzoate), which are subsequently degraded via the metacleavage pathway to form carboxylic acid pyruvate, and aldehydes. The genes of the upper pathway (designated *xyl*ABC) and the metacleavage pathway enzymes (designated *xyl*DEGF) are organized into two regulatory units controlled by the products of two regulatory genes (*xyl*S and *xyl*R).

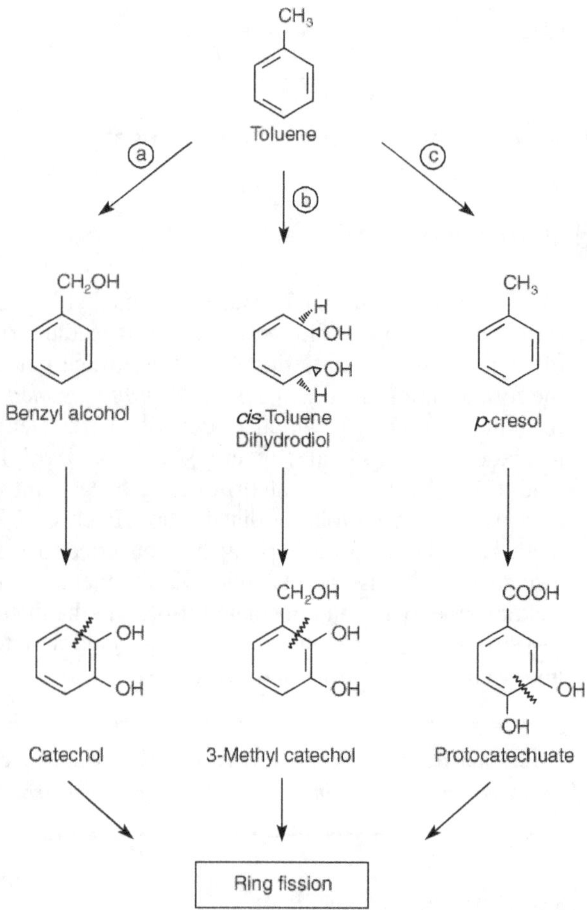

Figure 6.3 Toluene oxidation pathways to ring fission substrates a) *Pseudomonas putida* mt-2, b) *Pseudomonas putida* PpF-1, c) *Pseudomonas mendocina*

The enzymes for the complete degradation of toluene in Pseudomonas putida mt-2 are encoded by a set of genes borne on a TOL (for toluene) plasmid.

PHENOLS

Among the aromatics, the recalcitrant contaminant is phenol. It imparts objectionable taste and odour to the drinking water when it combines with chlorine to form chlorophenols. Phenols have high oxygen demand and in high concentration can deplete the oxygen of the receiving body of water. Oil refineries, chemical plants, explosive manufacturing, resin manufacturing and cake ovens are the sources for phenols, polyphenols, chlorophenols and phenoxy acids. Foundry operation, rubber reclamation plant, fibreglass manufacturing and textile mills are also the source for phenol contamination. Oil and cyanides are common in phenolic waste waters and they should be removed before removing phenol. Chemical oxidation, adsorption with activated carbon and biological methods are the routine procedures adopted for treating phenol-containing wastes.

Biological Degradation

Toxic substances must be removed from phenolic waste water by pre-treatment. Biological processes include lagoons, oxidation ditches, trickling filters and activated sludge. Activated sludge system is ideal because of its efficiency and ease of control. Organic and hydraulic loading fluctuations are the problems concentrated during biological processes. Spores of *Fusarium flocciferum* can be used to degrade phenol following immobilization on diatomic particles. Cell loading up to 50% dry weight have been obtained and the immobilized system can efficiently degrade a high concentration of substrate and is active for over two months.

Problems of Phenolic Degradation

Energy gains through aerobic degradation of phenolic compounds are high. However there are certain inherent characteristics of phenol, which auto-oxidizes into macromolecules in the presence of molecular oxygen or light. Such molecules hamper the progress of degradation and only monomers are attached by bacteria (Eismann *et al.,* 1996). In anaerobic condition, the auto-oxidation process is circumvented where the existing reducing conditions prevent the phenols from condensing into humus-like macromolecules. In many ecological niches, aerobic and anaerobic states often exist together since the solubility and penetration of oxygen is low (Table 6.2). Fuchs *et al.* (1994) have outlined the difference in aerobic and anaerobic degradation of phenolic compounds.

Table 6.2 Comparison of aerobic and anaerobic aromatic metabolism

Step	Aerobic	Anaerobic
Channelling	Oxygen	$+H_2O$, $+2(H)$, $-2(H)$, $+CO_2$, $+COH$, $+ATP$
Central intermediates	Catechol, protocatechuate, gentisate	Benzoyl CoA, resorcinol phloroglucinol
Properties of central intermediates	Easy to oxidize (cleave)	Easy to reduce (hydrate)
Attack at the ring	O_2	2 or 4(H) $+H_2O$
Ring cleavage	Oxygenolysis of aromatics	Hydrolysis of 3-oxocompound
Metabolic pathway	3-oxoadipate/TCA cycle	β-oxidation

POLYCHLORINATED BIPHENYLS (PCBs)

Polychlorinated biphenyls (PCBs) are mixtures of biphenyls with one to ten chlorine atoms per molecule. They are oily fluids with high boiling points, great chemical resistance, low electrical conductivity and high refractive index. PCBs, a group of 209 synthetic halogenated aromatic hydrocarbons, were first prepared in 1881, and since 1930 have been in general use in products that include heat transfer agents, lubricants, flame retardants, plasticizers, and waterproofing materials (Roberts *et al.,* 1978). Because of their wide range of physical properties, their chemical stability, and their miscibility with organic compounds, PCBs have also been used extensively as hydraulic fluids, adhesives, wax extenders, dedusting agents, and especially as dielectric fluids in capacitors and transformers. The number and position of the chlorine atoms on the biphenyl rings affect the biological properties of the compound. As a result of human activity, polychlorinated biphenyls (PCBs) are now distributed worldwide, with measurable concentrations reported in polar bears (*Ursus martinus*), birds, fish, wildlife, marine organisms of the Atlantic and Pacific Oceans, and up to 91% of the adult human populations. Their presence in the organisms has been shown to cause reproductive failure, birth defects, skin lesions, tumours, liver disorders and among sensitive species, death. PCB toxicity is further enhanced by their ability to bioaccumulate and biomagnify within the food chain due to extremely high liposolubility.

Sources

Environmental contamination resulted from industrial discharges, from leaks of supposedly closed systems, from disposal of PCB wastes to municipal sewage treatment plants, landfills, and equipment dumps, and especially through atmospheric transport of incompletely incinerated PCBs. Long-range atmospheric transport of PCBs by wind, rain, and snow is now well-documented (NAS, 1979). PCBs tend to bond tightly to particulate matter, notably soils and sediments of lakes, estuaries, and rivers, where they may remain available for re-suspension for at least 8 to 15 years (Swain, 1983). The North Atlantic Ocean seems to be the dominant sink for PCBs, accounting for 50 to 80% of the PCBs in the environment, while freshwater sediment is a major continental reservoir (NAS, 1979). Atmospheric deposition and high sediment contamination have also been implicated as major sources of PCB contamination (Rohrer *et al.,* 1982).

Nature of PCBs

PCBs are organic compounds commercially produced by chlorination of a biphenyl (BP) with anhydrous chlorine in the presence of iron filings or ferric chloride as the catalyst. The purified product is a complex mixture of chlorobiphenyls containing 18 to 79% chlorine; the precise composition depends on the conditions under which chlorination occurred (EPA 1980). Ten possible degrees of chlorination of the biphenyl molecule give rise to ten PCB congener groups: mono-, di-, tri-, tetra-, penta-, hexa-, hepta-, octa-, nano-, and decachlorobiphenyl. Within any congener group, a number of positional isomers (discrete chemical compounds) are possible, depending on the number of chlorines in the molecule. Although chlorine substitution is favoured at the *ortho-* and *para-* positions, the commercial products are complex mixtures of isomers and congeners with no apparent positional preference for halogen substitution (Safe, 1984). Recent advances in the identification and quantification of PCB isomers through mathematical and computer-assisted techniques (Schwartz *et al.*, 1984) will prove useful in data interpretation of metabolic rate studies. Toxic materials as impurities in PCBs include polychlorinated dibenzofurans (PCDF).

> *Polychlorinated biphenyls (PCBs) are now distributed worldwide, with measurable concentrations reported in polar bears (**Ursus martinus**), birds, fish, wildlife, marine organisms of the Atlantic and Pacific Oceans, and up to 91% of the adult human populations.*

Critical Levels of PCBs for Living Organisms

LC_{50} values of sensitive species of freshwater and marine organisms subjected to various Aroclor PCBs varied from 0.1 to 10.0 μg/l during exposure of 7 to 38 days. For aquatic life, water concentrations of less than 0.014 μg total PCBs/l (ppb) appear to afford a satisfactory degree of protection, although concentrations as low as 0.006 μg/l resulted in measurable accumulation by various species of filter-feeding shellfish. Among sensitive species of teleosts, total PCB residues (in μg/kg fresh weight) in excess of 500 in diets, 400 in whole body, and 300 in eggs were demonstrably harmful, and should be considered as presumptive evidence of significant PCB contamination. For all avian species, PCB residues of 310 mg/kg fresh weight or higher in brain were associated with an increased likelihood of death from PCB poisoning. For birds, total PCB levels (in μg/kg fresh weight) in excess of 3,000 in diet, 16,000 in egg, or 54,000 in brain were frequently associated with PCB poisoning. Among small mammals, the mink (*Mustela vison*) is one of the most susceptible species tested; dietary levels as low as 100 μg PCBs/kg fresh weight caused death and reproductive toxicity.

Effects of PCBs

PCBs elicit a variety of biological and toxic effects including skin lesions, a wasting syndrome, immunotoxicity, reproductive toxicity, genotoxic and epigenetic effects, hepatomegaly and related liver damage, and the induction of hepatic and extrahepatic drug-metabolizing enzymes. PCBs incorporated into phytoplankton exert inhibitory effects on photosynthesis and cell motility. In addition to direct toxic effects on algae, accumulated PCBs are readily introduced into the aquatic food chain (Rohrer *et al.,* 1982). Equilibrium levels of stable lipophilic contaminants in fish are directly proportional to the ambient water concentration of the chemical. In general, toxicity increased with increasing exposure, crustaceans and younger developmental stages were the most sensitive groups tested, and lower chlorinated biphenyls were more toxic than higher chlorinated biphenyls. PCBs were implicated in eggshell thinning and reproductive failure of some predatory and fish-eating birds. PCB accumulations from the diet and from other sources are high, and retention is lengthy in fatty tissues. In general, PCB accumulation is rapid and depuration is lengthy. Diet is an important route of PCB accumulation. The carcinogenic effects of PCBs have been established in mice and rats with various Aroclor and Kanechlor PCBs and these, in turn, may enhance the carcinogenicity of other chemicals (EPA 1980). Experimental data clearly shows that commercial PCBs cause liver damage which leads to putative preneoplastic changes and hepatocellular carcinomas; however, these lesions are observed only after lengthy (11 to 21 months) exposures to relatively high doses (100 to 1,200 ppm in diets) of these chemicals (Safe 1984).

Factors Affecting PCB Toxicity

The toxicological properties of individual PCBs are influenced primarily by two factors: the partition coefficient based on solubility in N-octanol/water (K_{ow}); and steric factors, resulting from different patterns of chlorine substitution. In general, PCB isomers with high K_{ow} values, and high numbers of substituted chlorines in adjacent positions, constitute the greatest environmental concern. Biological responses to individual isomers or mixtures vary widely, even among closely related taxonomic species.

> *PCBs incorporated into phytoplankton exert inhibitory effects on photosynthesis and cell motility.*

Microbial Degradation of PCBs

PCBs are extremely stable compounds, and are slow to chemically degrade under environmental conditions. Microbial degradation of PCBs depends on the degree of chlorination and the position of the chlorine atom on the biphenyl (BP) molecule; lower chlorinated BPs are readily transformed by bacteria, but not the higher chlorinated compounds (NAS 1979). Higher chlorobiphenyls, i.e., those with five or more chlorine atoms, were more persistent in the environment than those with three or less chlorine atoms; tetrachloro BPs were intermediate in persistence (EPA 1980). Bacterial degradation of PCBs has been shown by oxidative and reductive metabolic mechanisms. In the majority of bacteria, the gene specifying the oxidative catabolic pathways of PCBs include four enzymes: biphenyl dioxygenase (a gene product of *cbp*A), dihydrodiol dehydrogenase (a gene product of *cbp*B), 3-phenylcatechol dioxygenase (a gene product of *cbp*C), and 2-hydroxy-6-oxo-6-phenyl hexa-2,4-diconate (HOPDA) hydrolase (a gene product of *cbp*D). Analysis of detection of mutant plasmids has indicated that *cbp*ABCD genes are contained in an operon. An operon containing two *cbp*C and *cbp*D genes has also been reported in *Pseudomonas putida* and *P. alcaligenes.* Walia and Khan (1992) studied the expression of the PCB-degrading genes by biochemical and chemical analysis of 4-chlorobiphenyl (4-CBP) metabolites. They have cloned the genes of *Pseudomonas putida* capable of degrading PCBs in cosmid vector PCP13 and transferred into *P. putida* Ac812. The resulted bacteria containing recombinant cosmid pOH88 were able to degrade PCBs (Aroclor 1254).

Passage of PCBs through activated sludge in sewage treatment plants for 48 hours resulted in 81% degradation for Aroclor 1221 (21% chlorine by weight), 26% for Aroclor 1242 (42% chlorine), and only 15% for Aroclor 1254 (54% chlorine) (NAS 1979). Certain PCB congeners, such as 4-chlorobiphenyl, were highly mutagenic to *Salmonella typhimurium in* Ames tests (EPA 1980). Aroclor 1221 was less mutagenic, while Aroclors 1254 and 1268 were essentially inactive. In general, mutagenic activity tends to decrease with increasing chlorination (EPA 1980). Mutagenic properties of Aroclor 1221 and 4-monochlorobiphenyl to bacteria were indicated by positive Ames tests (NAS 1979).

> *Microbial degradation of PCBs depends on the degree of chlorination and the position of the chlorine atom on the biphenyl (BP) molecule.*

Biphenyls are able to serve some microbes as their only source of carbon and energy. Mono-, di-, tri- and tetrachlorobiphenyls are, to some degree, subject to biodegradation as evidenced by formation of hydroxylated metabolites (Safe, 1984). In extensively chlorinated biphenyls, the substituents prevent ring hydroxylation. The highly chlorinated analog cause the most concern with respect to persistence and environmental contamination. Chaudhry and Chapalamaduger (1991) reported several microbes that are capable of degrading PCBs. Anaerobic biodegradation of PCB is carried out by the white rot fungus *(Phanerochaete), Acinetobacter* and *Alcaligenes.* Degradation of PCB typically is by co-metabolism and is enhanced by the addition of less chlorinated analogs such as dichlorobiphenyl. The specific congeners are differently degraded. Various PCB products, according to their composition, exhibit different degrees of susceptibility to biodegradative transformations. More extensively chlorinated PCBs are more likely to be dechlorinated than the less chlorinated ones. *Meta-* and *para-*positions are dechlorinated in preference to *ortho-*positions.

> *The highly chlorinated analog cause the most concern with respect to persistence and environmental contamination.*

DEGRADATION OF DYES

Synthetic dyes are used extensively for textile dyeing, photography and as additives in petroleum products. The textile dyeing process requires a large volume of freshwater with fairly high purity and discharges equally large volume of waste water after the dyeing process. The waste water contains dyes at concentrations ranging from 10–200 mg/l and about 10–20% of the dye present in effluents along with organic and inorganic accessory chemicals involved in the dyeing process (Murugesan & Kalaichelvan, 2003). The presence of even a small fraction of dye in water (less than 1 ppm for some dyes) is highly visible due to the colour and affects the streams and other water bodies. Waste water from textile industry is a complex mixture of many polluting substances ranging from organochloride-based waste pesticides to heavy metals associated with dyes and the dyeing process (Correia, 1994). The untreated effluents released from the dyeing units cause a major threat to the environment. Around 10,000 different dyes with an annual production of more than 7×10^5 metric tonnes worldwide are commercially available and are extensively used in textile processing, paper, printing, pharmaceutical, food and other industries (Mc Mullan *et al.,* 2001). They are not readily degradable under natural conditions and are typically not removed from waste water by conventional waste-water treatment methods. It had been estimated that about 10% of the dyestuff used during the dyeing processes do not bind to fibres and are therefore released into the environment. The dyes/or their degradation products are toxic, mutagenic and carcinogenic in nature. The carcinogenicity may be due to the dye itself or to the aryl amine derivatives generated during the reductive biotransformation of the azo linkage (Zimmerman *et al.,* 1982). Azo dyes constitute more than 50% of the dyes produced annually and are the most important group of synthetic colourants that are extensively used in textile, food, pharmaceutical and printing industries. They are recalcitrant to biodegradative processes and pose toxicity to aquatic organisms and animals.

Classification of Dyes

A dye is a substance used to impart colour to fabric, food and other objects. Textile dyes are classified as azo-, diazo-, cationic-, basic-, anthraquinone-based and metal-complex based, depending on the nature of their chemical structure. They are also classified as reactive, disperse, vat mordant, etc. based on their application (Figure 6.4).

The presence of even a small fraction of dye in water (less than 1 ppm for some dyes) is highly visible due to the colour and affects the streams and other water bodies.

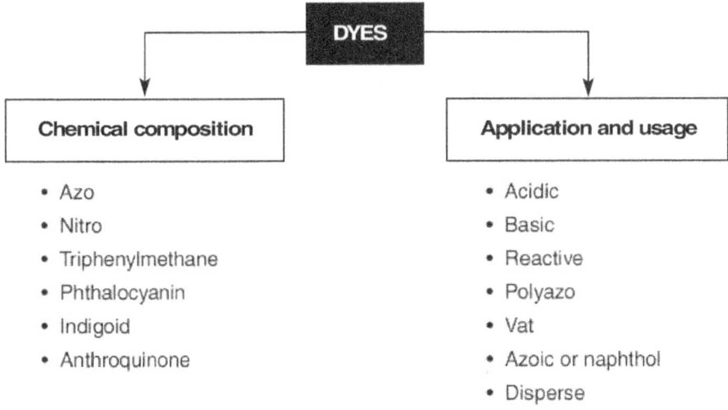

Figure 6.4   Classification of dyes

Azo dyes constitute more than 50% of the dyes produced annually and are the most important group of synthetic colourants that are extensively used in textile, food, pharmaceutical and printing industries.

Problems Caused by Dyes

The untreated effluents of textile dye industries may be highly coloured and thus particularly dangerous when discharged into the open water bodies. The following are the problems caused by these effluents.

1. Sunlight penetration of the stream would be reduced, which is essential for photosynthesis. Consequently, the ecosystem gets affected.

2. These effluents are toxic to fish and mammalian life.

3. They inhibit the activity and growth of microorganisms.

4. The colourants and their intermediates expose the organisms to chronic risks, and are carcinogenic and allergenic.

5. The azo dyes are capable of producing intestinal cancers.

6. Some dyes are teratogenic, capable of causing cerebral and skeletal abnormalities in foetuses (Murugesan and Kalaichelvan, 2003).

Impact of Azo Dyes

Azo dyes are the largest and most versatile class of dyes with one or more azo (N=N) bridges linking substituted aromatic structures. They are generally aromatic hydrocarbons, derivatives of benzene, toluene, naphthalene, phenol and aniline. A wide variety of azo dyes with anthraquinone, polycyclic and triphenylmethane are being used in textile dyeing and printing processes. Azo dyes have been linked to human bladder cancer, splenic sarcomas, hepatocarcinomas and nuclear anomalies in experimental animals and to chromosomal aberrations in mammalian cells. Azoreductases from bacteria are capable of converting some azo dyes to mutagenic and carcinogenic amines. Azo dyes induce liver nodules in experimental animals and there is a higher incidence of bladder cancer in dye workers exposed to large quantities of azo dyes. Boeniger (1980) reported that 1-aminonaphthalene (1-AN) forms the intermediate during the production of azo dyes, herbicides and rubber antioxidants. Benzidine-based azo dyes are widely used in dye manufacturing, textile dyeing, colour paper printing and leather industries. In 1980, the National Institute for Occupational Safety and Health published a survey of the data on the carcinogenicity of BZ-based dyes. Experimental studies with rats, dogs and hamsters have shown that animals administered BZ and BZ-congener-based dyes excrete potentially carcinogenic aromatic amines and their N-acetylated derivatives in their urine (Nony, and Bowman, 1980).

> *Azo dyes have been linked to human bladder cancer, splenic sarcomas, hepatocarcinomas and nuclear anomalies in experimental animals and to chromosomal aberrations in mammalian cells.*

Aromatic Amines

Aromatic amines are generally not degraded and they are accumulated under anaerobic conditions with the exception of few aromatic amines characterized by the presence of hydroxyl and/or carboxyl groups. Mineralization of the aromatic amines by aerobic sludge in treatment plants is more common. Some aromatic amines are readily auto-oxidized in the presence of O_2 to humic-like oligomeric and polymeric structures. The toxicity and carcinogenicity of certain azo dyes in mammalian systems may result either from interactions of the intact molecules with cytosolic receptors or from the formation of free radicals and acrylamines during azo reduction. Improved performance of the agitated cultures may be due to the increased mass and oxygen transfer between the cells and the medium. Glucose concentration has a strong effect on the decolorization of the dyestuff by the Basidiomycete strain F29. The decolorization ability decreases rapidly in the absence of glucose, supporting that the organism requires an additional carbon source to fuel the degradation process.

> *Mineralization of the aromatic amines by aerobic sludge in treatment plants is more common.*

Banned Amines

The aerobic biodegradation experiments revealed that aniline is readily degraded and the aerobic biodegradation of aniline is ubiquitous. Chloroanilines adversely affect the soil microflora (Gheewala, 1997). *P. acidovorans* CA28 and *Pseudomonas* sp. JL2 are capable of degrading chlorinated anilines and use them as the sole source of carbon. The prerequisite for the mineralization of many azo dyes is a combination of reductive and oxidative steps. The biodegradation of two azo dyes, 4-phenyl azophenol (4-PAP) and Mordant Yellow 10 are reduced, resulting in a temporal accumulation of aromatic amines. Further, 4-aminophenol (4-AP) and aniline are detected from the reduction of 4-PAP. Aniline is degraded further in the presence of oxygen by the facultative aerobic bacteria present in the anaerobic granular sludge. Aerobic enrichment cultures developed on aromatic amines combined with oxygen-tolerant anaerobic granular sludge can be used to completely biodegrade azo dyes under integrated anaerobic–aerobic conditions.

Impact of Dyeing Factory Effluent on Living Organisms

Intensive irrigation of agricultural lands with water polluted from various industrial effluents, severely affect soil fertility and plant growth. Dissolved substances in industrial effluents alter the chemical and biological status of the soil and water, which may affect growth and productivity of plants. Effluents also affect susceptibility of plants to various pathogens. Dyeing factory effluent that alters the colour and quality of the water bodies has been proved to be hazardous to aquatic organisms (Khan and Jain, 1995). A considerable decrease in the biochemical parameters of onion grown in untreated dye solution compared to the onion grown in treated dye solution was reported. Toxic compounds from dye effluent get into aquatic organisms, pass through food chain and ultimately reach man and cause various physiological disorders like hypertension, sporadic fever, renal damage, cramps, etc. Bioaccumulation of toxicants depends on availability and persistence of the contaminants in water, food and physico-chemical properties of the toxicants. Impact of the toxicants on fauna and flora had been reported (Amutha *et al.,* 2002). Naphthalene selectively accumulated in gills, liver, gut and gall bladder of the mud-sucker, *Glichthus mirabilis,* nose-kill fish, *Fundulus similis* and have shown accumulation of naphthalene in the brain. At lethal concentration, naphthalene primarily affects blood component while at sublethal concentrations, it causes neurosensory damage and metabolic stress (Dimichele and Taylor, 1978).

Treatment of Dye Effluents

Treatment of waste water containing dyes and decolorization is very difficult due to wide range of pH, salt concentration and chemicals. The majority of colour removal techniques work either by concentrating the colour into a sludge, or by partial to complete breakdown of the coloured molecule. Effluent treatment methods may be classified into three main categories: physical, chemical and biological (Figure 6.5). The physical and chemical methods are limited due to excessive use of chemicals, sludge generation with subsequent disposal problems, high installation as well as operating costs and sensitivity to a variable waste-water input.

> *Aerobic enrichment cultures developed on aromatic amines combined with oxygen-tolerant anaerobic granular sludge can be used to completely biodegrade azo dyes under integrated anaerobic–aerobic conditions.*

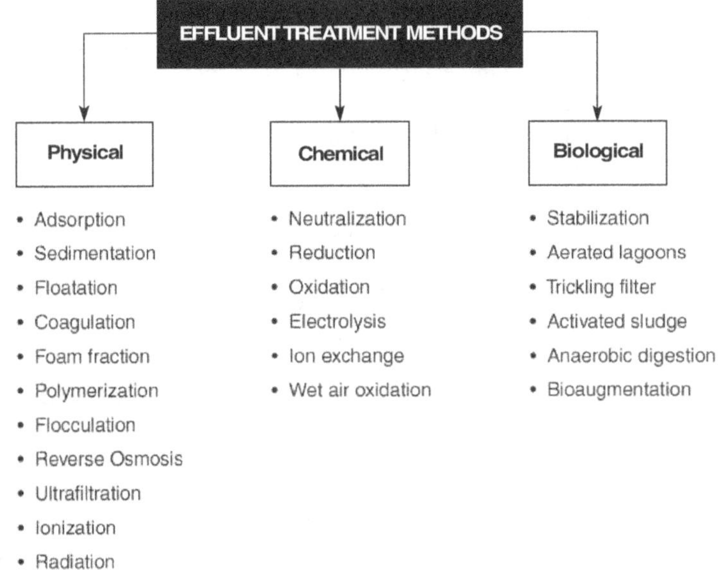

Figure 6.5 Effluent treatment strategies

Bioremediation of Dyes

Bioremediation constitutes an alternative to conventional physico-chemical methods. Biological systems are recognized by their capacity to reduce biological and chemical oxygen demand by aerobic biodegradation. Biological treatment of textile effluents may be either aerobic, anaerobic or a combination of both (Figure 6.6) depending on the type of microbe being employed (Keharia and Madamwar, 2004).

Constraints in Dye Degradation

Dye decolorization can be performed by using one or a combination of adsorption, filtration, precipitation, chemical degradation, photodegradation and biodegradation. However, the colour and chemical composition of textile effluents are usually subjected to daily and seasonal variations dictated by production routines and fashion cycles. Dyes are highly resistant to microbial degradation under aerobic conditions in waste-water treatment plants, since they are designed to resist chemical fading and light-induced oxidative fading (Nigam *et al.,* 2000). Other factors involved in reduced biological degradation include properties, such as high water solubility, high molecular weight and fused aromatic ring structures, which inhibit permeation through biological membranes. Substitutions such as azo, nitro or sulphur groups in dyes are responsible for recalcitrance.

Among reactive dyes, azo dyes are problematic due to their excess consumption and water solubility. Sulphonic acid and the azo group confer xenobiotic character to the azo dyes. Anthraquinone-based dyes, due to their fused aromatic structures, remain coloured for long periods. Brilliance and higher colour intensity of basic dyes make them more difficult to decolorize. As the dyestuffs are recalcitrant to biodegradation, they escape various stages of waste-water treatment plants and enter the environment. A huge amount of sludge is formed during effluent treatment process by chemical precipitation method. This sludge is toxic and highly problematic to safe disposal. The detoxification and disposal of the sludge is a major problem faced by the textile dye units.

> *The majority of colour removal techniques work either by concentrating the colour into a sludge, or by partial to complete breakdown of the coloured molecule.*

Environmental Factors on Dye Degradation

Culture conditions affect the physiology, expression and the activity of microbes involved in degradation. Culture condition optimization is essential for effective dye degradation. The most important external factors affecting the activity of the organisms are temperature, pH, dissolved oxygen concentration and fixed nitrogen concentration. pH had been reported as one of the factors that affects dye decolorization. Agitation also plays a vital role in decolorization. The superior and increased performance of agitated culture may be due to the physiological state of the microbe and increased mass and oxygen transfer between the cells and the medium due to mixing (Murugesan and Kalaichelven, 2003). The formation of a mat at the surface in static cultures restricts O_2 transfer to the cells beneath the surface and in the medium. Oxygen limitation is likely to inhibit the oxidative enzymes and prevent decolorization. Glucose concentrations have a strong effect on the decolorization of the dyestuff by the fungi. The nature and the quantity of the available nitrogen sources exert a great influence on the extracellular lignolytic enzyme production and decolorization activity of wood-rotting basidiomycetes. The optimum conditions differ from one species to the other. Hence optimization has to be done with regard to the aim of the application.

> *Dyes are highly resistant to microbial degradation under aerobic conditions in waste-water treatment plants, since they are designed to resist chemical fading and light-induced oxidative fading.*

Dye Decolorization Mechanisms

Dye decolorization involves two steps:

1. physical adsorption and
2. enzymatic degradation (Figure 6.6).

Binding of dyes to the fungal hyphae and physical adsorption and enzymatic degradation by extracellular and intracellular enzymes are the reasons for the colour removal (Young and Yu, 1997). The dye-saturated mycelium can be regenerated and used for repeated dye adsorption. Enzymes such as lignin peroxidase (LiP), manganese-dependent peroxidase (MnP) and laccase involved in lignin degradation also participate in dye decolorization. Extracellular fluid from cultures of *P. chrysosporium* and purified lignin peroxidase were able to degrade crystal violet and six other triphenyl methane dyes by sequential N-demethylation. The mechanism of azo dye oxidation by peroxidases such as lignin peroxidase probably involves the oxidation of the phenolic groups to produce a radical at the carbon bearing the azo linkage. Then water attacks the phenolic carbon to cleave the molecule producing phenyldiazine. The phenyldiazine can be oxidized by a one-electron reaction generation (Chivukula *et al.*, 1995). Further degradation of azo dyes involves aromatic cleavage which has also been found to be dependent on the identity of the ring substituents with the presence of phenolic, amino, acetamino, 2-methoxyphenol or other easily biodegradable functional groups resulting in a greater extent of degradation.

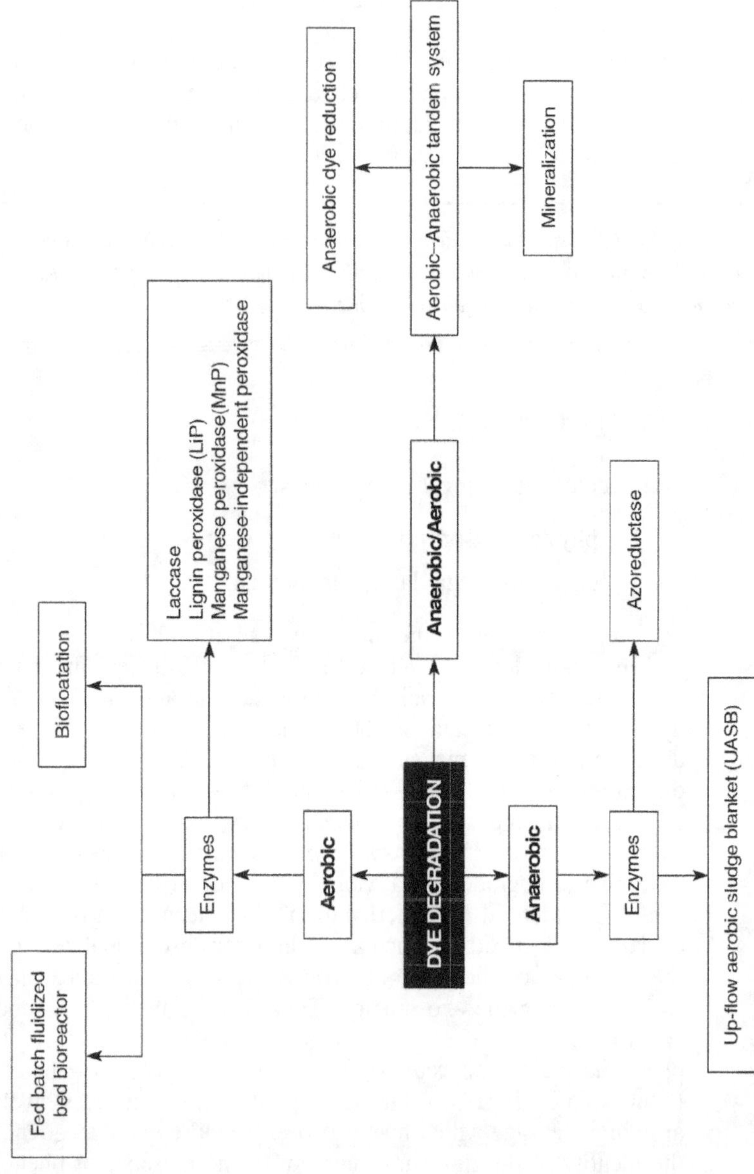

Figure 6.6 Various dye treatment strategies

> *The detoxification and disposal of the sludge is a major problem faced by the textile dye units.*

Role of Azoreductase in Biodegradation

Azoreductase of microorganisms are favourable for the development of biodegradation systems for azo dyes, because these enzymes catalyse reductive cleavage of azo groups (—N=N—) under mild conditions. The most generally accepted hypothesis is that many bacterial strains possess rather unspecific cytoplasmic enzymes, which act as 'azoreductases' and under anaerobic conditions transfer electrons via soluble flavins to the azo dyes. Azoreductase catalyses the reductive cleavage of azo linkages in benzidine-based dyes and other compounds containing an azo bond to produce aromatic amines. Decolorization and a decrease in the absorbance of the dye accompany this reduction step. Azoreductases have been detected in liver cells and several anaerobic bacteria. The intestinal microbial flora of humans, monkeys and rats contain bacteria that reduce the azo dye to aromatic amines. The azoreductases of these bacteria are similar in their functions, antigenicity stimulation by flavin adenine dinucleotide and oxygen sensitivity and have structural similarity. The recombinant strains expressing high level of flavin reductase in the presence of oxygen, is assumed as azoreductase activity, since it react rapidly with molecular oxygen. It was found that under aerobic conditions there may be some decrease in the concentration of dye. Almost identical oxidation rates for NADH were observed under aerobic and anaerobic conditions. Azo reductases from several bacteria including *Clostridium perfringens* and human intestinal anaerobic bacteria, catalyse the reductive cleavage of azo linkages in azo dyes to produce aromatic amines.

When the azo dye was incubated in oxygen-free buffer with NADH as a source of reduction equivalents, a slow decolorization of the azo dye was observed. Addition of cell extract in the presence of flavin adenine dinucleotide (FAD) to the reaction mixture clearly enhanced the reaction rate drastically after a few minutes and shortened the lag phase. A possible explanation for this pronounced effect is that FAD is reduced enzymatically by NADH, and $FADH_2$ can then spontaneously reduce the azo dye. Cofactors like $FADH_2$,

FMNH$_2$, NADH and NADPH, as well as the enzymes reducing these cofactors are located in the cytoplasm. Lysis of cells would release cofactors in the extracellular environment. Hence, it has been reported several times that cell extracts or starving or lysed cells show higher azo dye reduction rates than intact or resting cells. However, for intact cells, a membrane transport system would be a prerequisite for the reduction of azo dyes by these cofactors. This presents a serious obstacle, especially for dyes containing sulphonate groups. In addition, FAD and FMN cannot readily cross cell walls. In contrast, riboflavin is able to move across cell membranes. *Bacillus* sp. OY12 transforms azo dyes into colourless compounds, and this reduction is mediated by azoreductase activity in the presence of NADPH. A gene encoding azoreductase of *Bacillus* sp. have been cloned. Similarly, azoreductase gene was cloned from *Xenophilus azovorans* KF46F, grown in the carboxylated azocompound 1-(4-carboxy phenyl azo)-2-naphthol as the sole source of carbon and energy. A presumed NAD(P)H-binding site was identified in the amino terminal region of the azoreductases. An FMN-dependent NADH- azoreductase of *E. coli* was purified and the N-terminal sequence of the azoreductase was confirmed. The gene for ACP was amplified using genomic DNA of *E. coli*. Over-expression of the *acpD* gene provided the *E. coli* with the large amount of 23 kDa protein and more than 800 times higher azoreductase activity. Its importance has been deduced from the wide distribution of the gene in other bacteria, such as *Azospirillum brasilens, Bacillus subtilus, B. stereothermophilus, Pseudomonas aureginosa* and *Mycoplasma pneumoniae.*

Microbial Degradation of Dyes

The textile industry is a major user of water, starting from washing raw wool or man-made fibre production up to garment manufacturing. The ability of microorganisms to decolorize and metabolize dyes has long been known and the use of bioremediation-based technologies for treating textile waste water has attracted interest (McMullan *et al.,* 2001). Several microorganisms have been found to decolorize and mineralize azo dyes (Table 6.3). The bacterial metabolism of azo dyes is initiated by a reductive cleavage of the azo bond in most cases, which results in the formation of amines. These reductive processes have been studied in some aerobic bacteria, which grow on azo compounds. The sulphonated amines that are formed

in the course of these reactions may be degraded aerobically
(Stolz, 2001). Two-stage anaerobic treatment process is adapted to
overcome the problem of recalcitrance of azo dyes. The azo dye is
readily reduced to the corresponding colourless aromatic amines in
the first stage and then metabolized. Under anaerobic conditions
the bacterial consortium reduced, the sulphonated azo dye. Mordant

Table 6.3   Microbes involved in dye degradation

Microbe	Name of the Dyes
Phanerochaete chrysosporium	Azo, Anthraquinone, Triphenylmethane Orange II, Tropeolin
Geotrichum candidum	Complex azo dyes
Trametes versicolor	PCBs
T. pocas, T. cingulata	Anthracene
Pleurotus ostreatus	Fluorene
Clitrocybulla dusenii	Phenanthrene benzo(α)pyrene
Bjerkandera adusta	Dichloroaniline
Pycnoporus cinnabarinus	HRB 8
Datronia concentria	Heterocyclic and polymeric
Citrobacter sp., *Kurthia* sp. *Corynebacterium, Mycobacterium* sp. *Pseudomonas mendocina* and *P. alcaligenes*	Triphenyl methane
Candida zeylanoides,	Azo dyes
Aeromonas sp., *Bacillus subtilis, Proteus mirabilis, P. luteola, Pseudomonas pseudomallei*	Reactive Red 22
Pyricularia oryzae	Phenol azo dyes

> *Two-stage anaerobic treatment process is adapted to overcome the problem of recalcitrance of azo dyes.*

Yellow 3 and the aromatic amines, 6-aminonaphthalene-2-sulphonate and 5-aminosalicylate were generated and excreted into the medium. The amines were mineralized after aeration of the culture. The ability of the mixed bacterial culture to reduce the azo dye possessed the ability to oxidize various naphthalene sulphonic acids (O'Neill *et al.,* 2000). Sulphonated azo compounds are widely used as dyes for textile industries. The aromatic compounds carrying SO_3H group as substituents often resist biodegradation and are not degraded under anaerobic condition (Bumpus, 1995). Certain carboxylated analogs of sulphonated azo compounds are utilized aerobically as the sole source of carbon and energy.

Microbiological treatment of sludge is the best way for the detoxification. White rot edible mushrooms grow efficiently on sludge and produce high amount of laccase and reduce the toxicity of the sludge (Murugesan, 2002). Membrane transport of the dyes seems to be the rate-limiting step in the microbial reduction of azo dyes. In most cases, the dyes adsorbed by the cell walls are reduced at slow rate.

A mixed culture of *Pseudomonas aeruginosa* 3MT and *Pseudomonas* sp. CP4 simultaneously and efficiently degrade mixtures of 3-chlorobenzoate (3-CBA) and phenol/cresols. Strains 3MT and CP4 used *ortho-* and *meta-* ring cleavage pathways respectively. The microbial degradation of sulphonated aromatic compounds is often accomplished by mixed cultures. *Pseudomonas* sp. CPE1 found to be capable of completely mineralizing 4-chlorobiphenyl via 4-chlorobenzoate. Co-cultures of three bacterial strains mineralized both chlorinated biphenyl and polychlorinated biphenyl. The microbial degradation was hindered by polyaromatic hydrocarbon, chloroanilines and chlorobenzenes due to adsorption of the substance to clay and soil organic matter. *Pseudomonas* sp. CPE1 capable of co-metabolizing low chlorinated biphenyls exhibited both 4-CB and 4-CBA biodegradation phenotypes. This strain was capable of using 1-CB as the sole carbon source and of completely dechlorinating 0.5 g/l of this xenobiotic after 3 days of incubation.

Aerobic Treatment Processes

The higher stability of dyes to microbial oxidation and their poor adsorption onto the activated sewage sludge are the main constraints of aerobic biological treatment (Greaves *et al.,* 1999). Majority of white-rot fungi have the ability to degrade a wide variety of dyes. The dye-decolorizing activity of these fungi has been correlated to their ability to produce lignolytic enzymes, such as laccase, lignin peroxidase, manganese peroxidase and manganese independent peroxidases, which exhibit broad substrate specificity. *Phanerochaete chrysosporium* has been shown to degrade a large spectrum of azo dyes with decolorization efficiencies exceeding 90% in most cases. Dyes having hydroxyl, amino, acetamino or nitrofunctional groups substituted in their aromatic ring are more susceptible to degradation by *P.chrysosporium*. Among various reactor configurations, a fed-batch-fluidized bed bioreactor was found to be most efficient, and showed 97% colour removal at a dye (Orange II) loading rate of 1000 mg/l per day. The reactor performance was highly stable when operated continuously, with more than 95% dye removal efficiency at dye loading rates of 100 mg /lper day for 30 days. Immobilized laccase can reduce 80% dye toxicity and the effluent could be successfully reused for dyeing.

> *The initial process in the microbial degradation of azo dyes is the cleavage of highly electrophilic azo bond leading to decolorization of azo dyes.*

The initial process in the microbial degradation of azo dyes is the cleavage of highly electrophilic azo bond leading to decolorization of azo dyes. The reductive cleavage of the azo bond resulted in the formation of aromatic amines as end products. However, most such end products of anaerobic degradation of azo compounds are further degraded by aerobes (Nortemann *et al.,* 1986). Aerobic bacteria degrade certain aromatic amines and sulphonated amino aromatics. Aromatic amines with one or more OH group tend to auto-oxidize in the presence of oxygen. This auto-oxidation process was observed for 4-aminophenol (4-AP) as well as for 5-amino salicylic acid (5-ASA). Cleavage of azo bond generates aromatic amines that are not degradable under anaerobic condition. But aerobic treatment can mineralize aromatic amines by non-specific enzymes through hydroxylation and ring cleavage. Components such as 6-amino naphthalene-2-sulphonic acid have been shown to be degraded by mixed aerobic cultures. A range of aromatic amines are aerobically biodegraded rapidly and hence they are unlikely to remain in the environment for a long time. Therefore, anaerobic treatment followed by aerobic treatment can be efficiently used to decompose putatively toxic and carcinogenic compounds (Feigel and Feigel, 1993).

A novel technology known as biofloatation has proved to be successful in the treatment of effluents from dyeing and finishing processes. Biofloatation leads to the transformation of tensioactive compounds present in waste water into activated foam that acts as a supporting structure with a high surface area for bacterial activity. The efficiency of the system is based on the breathing capacity of the bacterial pool working within this activated foam, which transforms organic waste into carbon dioxide and water at a faster rate than conventional systems. Due to its simplicity of use and low sludge production, the biofloatation process is ideal for treating waste waters from dyeing and finishing operations.

Anaerobic Treatment Processes

Anaerobic treatment is advantageous because of its simplicity and low running cost. The process could be used to decolorize dye-containing waste waters and for waste water biodegradability for subsequent aerobic treatment. Fed-batch bioprocess using *Pseudomonas luteola* was shown to effectively decolorize reactive Red 22. Continuous feeding of dye (200 mg/h) to a 2-l initial culture-volume and a yeast-extract to-dye ratio (Y/D) of 1.25 were found to be optimum conditions for dye decolorization, with a specific decolorization rate of 113.7 g/cell per hour. The Y/D ratio is critical for dye decolorization and a Y/D ratio higher than 0.5 is recommended for efficient decolorization. A pilot-scale two-stage, up-flow anaerobic sludge blanket bioreactor, consisting of an acidification tank and a UASB (Up-flow aerobic sludge blanket) reactor with a hydraulic retention time of about 12 hours for each phase, effectively decolorizes three different reactive dye bath effluents using tapioca (50 mg/l) as a co-substrate. Anaerobic methanogenic bacteria are responsible for colour reduction. Combination of dye adsorption on cellulosic anion-exchange resin and anaerobic dye reduction is effective in the removal of anionic dyes from waste water. Under anaerobic conditions, bacteria mediate decolorization of azo dyes by causing reductive cleavage of azo linkage. The process of azo bond reduction is catalysed by a variety of soluble cytoplasmic enzymes with low substrate specificities known as azoreductases. These azoreductases facilitate the transfer of electrons via soluble flavins or other redox mediators to the azo dye, which is then reduced. Reduced flavins that are the products of cytoplasmic flavin reductases may function as intermediates in azo dye reduction. Reduced organic compounds, such as Fe^{2+}, H_2S, etc., which are formed extracellularly as end products of certain strictly anaerobic bacterial metabolisms, may cause azo dye reduction. The anaerobic process offers several advantages in comparison to aerobic ones in terms of no aeration requirements, low sludge generation and methane gas production under anaerobic condition which leads to the formation of colourless aromatic amines, which are generally recalcitrant to further anaerobic degradation (Manu and Chaudhari, 2000).

> *A novel technology known as biofloatation has proved to be successful in the treatment of effluents from dyeing and finishing processes.*

Anaerobic/Aerobic Treatment of Dyes

Aerobic treatment after anaerobic dye reduction would be preferable for complete mineralization of dyestuffs in textile waste waters. Anaerobic reduction of azo dyes is relatively easy to achieve; complete mineralization is difficult and generally results in accumulation of aromatic amines. Aerobic bacteria mineralize these aromatic amines. Co-cultivation of oxygen-tolerant, anaerobic granular sludge with aromatic amine-degrading aerobic enrichment cultures, demonstrated mineralization of azo dyes, Mordant Yellow 10, and 4-phenyl azo phenol in the presence of ethanol as co-substrate. A sequencing batch reactor namely "an aerobic–anaerobic" tandem system was used to decolorize synthetic dye waste water containing reductive turquoise blue (RTB). The aerobic and anaerobic phases were performed separately in two reactors. Biodegradability of streams having a BOD_5 and a COD of less than 0.5 can be increased by Fenton's oxidation pretreatment and activated carbon. The resulting effluent could then be effectively treated with fixed film bioreactor, acclimatized with dyestuff effluent. Chemical oxidation, activated carbon adsorption and fixed bed biofilm processes with optional Fenton oxidation as a post-treatment system seems to be effective in treating dyestuff waste water having high salinity, colour and non-biodegradable organic content. Unhydrolysed Remazol Black B can be removed by aerobic processes. Dye removal increased with an increase in the biomass, whereas a high dye-to-biomass ratio inhibited colour removal. Hydrolysed dyes can be readily reduced within 24 hours under anaerobic conditions.

A sequential anaerobic and aerobic treatment process based on a mixed culture of bacteria isolated from textile-dye-effluent-contaminated soil was used to degrade sulphonated azo dyes Orange-G, Amido Black, Direct Red. Under anaerobic conditions in a fixed bed column using glucose as co-substrate, the azo dyes are reduced and the bacterial cells released the amines. The amines were completely mineralized in a subsequent aerobic biological effluent treatment involving both adsorption to cell biomass and degradation through anaerobic reduction. Degradation releases aromatic amines that are toxic and recalcitrant to anaerobic treatment but degradable aerobically. Many dyes used in textile industry cannot be degraded aerobically as the enzymes involved in aerobic degradation are dye-specific (Kulla, 1981).

Immobilization for Dye Decolorization

Immobilization is a practical treatment system for dye degradation. Microbes that are capable of dye degradation can be immobilized in carriers such as sintered glass, nylon web, polyurethane foam, porous polystyrene and silicon tubing. The immobilization method has several advantages over other methods, since it is easy to perform under aseptic condition *in situ* within the bioreactor and easy to scale up. During the last few years, different reactor designs have been proposed for an effective continuous anaerobic/aerobic treatment of azo dyes. They are fixed film bioreactors, packed bed bioreactors, anaerobic/aerobic rotating biological contactors, aerobic suspended-bed activated sludge reactor, anaerobic up-flow fixed bed column together with anaerobic agitated tank and pulse flow bioreactors (Kapdan *et al.,* 2002). Immobilization prevents cell washouts and allows a high cell density to be maintained in a continuous reactor. Since the catalytic stability is often improved by immobilization, microorganisms may tolerate and degrade higher concentrations of toxic compounds than their free cell counterparts. It has been suggested that immobilization results in enhanced substrate uptake because of the increased nutrient availability at the solid–liquid surface or the damage to the cell wall facilitating entry of substrate (D'Souza, 1980). Slight agitation (120 rpm) of the beads favoured rapid degradation, presumably caused by increased O_2

> *Many dyes used in textile industry cannot be degraded aerobically as the enzymes involved in aerobic degradation are dye-specific.*

supply to the beads and translocation of substrates and products. The rate of diffusion of substrate across the bead appears to limit its utilization, and gentle agitation accelerated the reaction.

Conclusion

A single step treatment process would be ineffective for the treatment of dyestuff industry waste water and a combination of more than one process may be useful. This is due to composition variation of dyes from industries, which may use more than one dye in a single dyeing operation in order to achieve designed colour shades. In general sequential anaerobic- aerobic process seems to be more practicable. Genetic modification of the enzymatic systems of different microbes responsible for decolorization and degradation may be useful in making the system more efficient.

BIOCONVERSION OF SPECIFIC POLLUTANTS

7

Heavy Metal Remediation
Pesticide Degradation
Synthetic Polymers
Dioxins
Radioactive Wastes

GET RID OF MALARIA, BUT INVITE PLAGUE!

In the mid-twentieth century, malaria was rampant in the Indonesian island of Sabah, earlier known as North Borneo. In 1955, the WHO began spraying the island with dieldrin (chemical related to DDT) to kill mosquitoes. The attempt was successful and malaria was almost eradicated. Dieldrin killed many other insects including flies and cockroaches. The lizards ate these insects and they died too. So did the cats that ate the lizards. Once the cats declined, rats proliferated in huge numbers and there was the threat of plague. The WHO then dropped healthy cats on the island by parachute. Dieldrin had also killed wasps and other insects that consumed a particular caterpillar, that was somehow not affected by the chemical. The caterpillars flourished and ate away all the leaves in the thatched roofs of the houses and the roofs started caving in. Ultimately the situation was brought under control.

Miller (2004)

HEAVY METAL REMEDIATION

The growing industrialization and modern agricultural practices that have spread worldwide have adversely affected the ecosystem. These practices leave persistent toxic metals and organic pollutants in the surroundings, which tend to accumulate and deteriorate the environment (Ahmed *et al.*, 2004). Contamination of heavy metals in the environment is a major global concern because of their toxicity and threat to human life and environment. Much research work has been carried out on heavy metal contamination in soils from various anthropogenic sources such as industrial wastes, automobile emissions, mining activity and agricultural practices. The recognition of toxic effects from minute concentrations of some metal ions has resulted in regulation of laws to reduce their presence in the environment to very low levels as ppb (i.e., μg l^{-1}). The number of heavy metals is about 65 and these are defined with respect to a number of criteria such as their cationic-hydroxide formation, specific gravity greater than 5 g/ml, complex formation, hard-soft acids and bases, and, more recently, association with eutrophication and environmental toxicity.

Roane and Pepper (2000) classified metals into three classes on the basis of their biological functions and effects. They include

1. The essential metals with known biological functions, (Na, K, Mg, Ca, V, Mn, Fe, Co, Ni, Cu, Zn, Mo, and W).

2. The toxic metals (Ag, Cd, Sn, Au, Hg, Ti, Pb, Al, and metalloids Ge, As, Sb and Se).

3. The non-essential, non-toxic metals with no known biological effects (Rb, Cs, Sr, and T).

> *Contamination of heavy metals in the environment is a major global concern because of their toxicity and threat to human life and environment.*

Based on primary accumulation mechanisms in sediments, heavy metals are classified into five categories.

1. adsorptive and exchangeable
2. bound to carbonate phases
3. bound to reducible phases (Fe and Mn oxides)
4. bound to organic matter and sulphides
5. detrital or lattice metals

Need for Metal Bioremediation

Greater awareness of the ecological effects of toxic metals and their biomagnifications through the food chain as well as highly publicized episodes such as mercury pollution in Minamata, Japan, have prompted a demand for decontamination of heavy metals in the aquatic systems. However, essential metals are required for enzyme catalysis, nutrient transport, protein structure, charge neutralization and control of osmotic pressure. Olson *et al.* (2001) reported incorporation of nickel in four microbial enzymes involved in ureolysis, hydrogen metabolism, methane biogenesis and acetogenesis. A number of heavy metals are required as micronutrients to plants. They act as cofactors as part of prosthetic groups of enzymes which are involved in a wide variety of metabolic pathways. However, when they are present in high levels, most heavy metals are toxic to plants.

Table 7.1 Application of heavy metals and their toxic effects

Metal	Use	Effect
Hg	Coal, vinyl chlorides, electrical batteries	Brain damage often leading to death
Pb	Plastic, paint, pipe, batteries, gasoline, auto exhaust	Neurotoxic
Ar	Pesticides, coal, detergents	Liver cirrohosis, mental disturbance, cancer, ulcer and hypokerotosis
Cd	Fertilizers, plastic, pigments	Kidney damage, injury in CNS and mental retardation
Cr	Tanning, paints, pigment, fungicide	Nephritis, cancer and ulceration
Zn	Fertilizer	Vomiting, renal damage and cramps
Co	Vitamin B-12	Diarrhoea, low blood pressure and paralysis
Se	Coal, sulphur	Damage of liver, kidney, spleen and nervousness
Ni	Electroplating	Teratogenic, carcinogenic, genotoxic and mutagen
Be	Coal, rocket fuel	Carcinogen, acute and chronic poison

> *A number of heavy metals are required as micronutrients to plants. They act as cofactors as part of prosthetic groups of enzymes which are involved in a wide variety of metabolic pathways.*

Metal concentration has been linked to birth defects, cancer, skin lesions, retardation leading to disabilities, liver and kidney damage and a host of other maladies (Table 7.1). The Center for Disease Control (CDC) and the Agency for Toxic Substances and Disease Registry (ATSDR) estimate that 15–20 % of U.S. children have lead levels greater than 15 mg/dl in blood, which is considered potentially toxic. Changes in the trace-element profile of the soil causes physiological and genetic changes in various forms of life, such as plants, aquatic and benthic fauna, insects, earthworms, fish, birds and mammals as evidenced by recent research work. Tyagi *et al.* (1999) stated that the concentration of heavy metals in industrial areas in India is much higher than the permissible limit of the World Health Organization. They have also reported that these metals in the ground water have caused various diseases in human beings and also disturb their metabolic functions.

> *Changes in the trace-element profile of the soil causes physiological and genetic changes in various forms of life, such as plants, aquatic and benthic fauna, insects, earthworms, fish, birds and mammals.*

Metal Availability

Sposito (2000) categorized metals as bioavailable (precipitated, non-sorbed and mobile) and non-bioavailable (precipitated, sorbed, and non-mobile) metals. In nature, metals and metalloids exist mostly as cations, oxyanions, or both in aqueous solution and mostly as salts or oxides in crystalline (mineral) form or as amorphous precipitates in soluble form. The mobility of metals as hydrated ionic salts is dependent on two factors.

a. The metallic element that is participating as positively charged ions (cations)

b. The one, which makes up the negatively charged component of the salt. The cationic/anionic solubility relationship is

> *In nature, metals and metalloids exist mostly as cations, oxyanions, or both in aqueous solution and mostly as salts or oxides in crystalline (mineral) form or as amorphous precipitates in soluble form.*

given in Table 7.2. Geochemical forms of heavy metals in soil affect their solubility, which directly influences their bioavailability. Lena and Rao (1997) reported that the mobility and bioavailability of metals in soils is in the order of $Zn > Cu > Cd > Ni$, based on their solubility and geochemical forms. The fate of toxic metals in soils depends mainly on the initial chemical form of the metal even though the environmental and edaphic conditions such as pH, redox status, and soil organic matter content have significant influence. Studies of Sauerback and Rietz (1983) have indicated the existence of different binding forms and strongly pH-dependent solubility effects of the trace elements. Soil cation exchange capacity (CEC) affects metal bioavailabilty and it depends on the organic matter and clay content of the soil. The toxicity of metals within soils with high CEC

Table 7.2 Summary of cation/anion solubility relationship of metals

Anions	Cations	Solubility
All anions	Na^+, K^+, NH_4^+	Soluble
NO_3^-, NO_2^-, $C_2H_3O_2$, MnO_4, ClO_4^-, ClO_3^-	All metals	Soluble
Cl^-, Br^-, I^-	All metals except Pb^{2+}, Ag^+, Hg^{2+}	Soluble
SO_4^{2-}	All metals except Ba^{2+}, Sr^{2+}, Pb^{2+}	Soluble
O^{2-}, S^{2-}, OH^-	All metals except Ca^{2+}, Ba^{2+}, Sr^{2+}	Insoluble
CO_3^{2-}, PO_4^{3-}, SO_3^{2-}, BO_3^{3-}, F^-, SiO_3^{2-}	All metallic salts except Na^+, K^+, NH_4^+	Insoluble

is low even at high total metal concentration and in contrast, the toxicity is vice-versa at low CEC. Under oxidized or aerobic conditions (+ 800 to 0 mV), metals are usually found as soluble cationic forms (Cu^{2+}, Cd^{2+}, Pb^{2+} and Ca^{2+}) and in reduced or anaerobic conditions (0 to –400 mV) they are found as CuS, PbS, and $CdCO_3$ precipitates. At low soil pH, the metal bioavailability increases due to its free ionic species and in contrast, high soil pH decreases due to insoluble metal mineral phosphate and carbonate formation. However, in contrast to other metal bioavailabilities, at high pH range of 5–9, nickel may be adsorbed on iron and manganese oxides, or form complexes with inorganic ligands (OH^-, SO_4^{2-}, Cl^- or NH_3).

> *Soil cation exchange capacity (CEC) affects metal bioavailability and it depends on the organic matter and clay content of the soil.*

Metal–microbe Interaction

Microbial interactions with heavy metals were reviewed by, Ehrlich (1997), Beveridge and Doyle (1997). Microbial interactions with small quantities of metals or metalloids do not exert a major impact on metal or metalloid distribution in the environment, whereas interactions with larger quantities are required in energy metabolism, for instance, to have noticeable impact. However, due to strong ionic nature metals bind to many cellular ligands and displace native essential metals from their normal binding sites, which are toxic (Figure 7.1). Eubacteria and archaea are able to oxidize Mn (II), Fe (II), Co (II), Cu(I), AsO_2^-, AsO, Se^o or reduce Mn (IV), Fe (III), Co (III), AsO_4^{2-}, SeO_3^{2-} on a large scale and conserve energy in these reactions. Some microbes reduce ions such as Hg^{2+} or Ag^+ to Hg^o and Ag^o respectively, but do not conserve energy from these reactions. Prokaryotes methylate metal and metalloid compounds producing corresponding volatile metal derivatives. The bacterial oxidation of AsO_2^- to AsO_4^{3-} by a strain of *Alcaligenes faecalis* and reduction of CrO_4 to $Cr(OH)_3$ by *Pseudomonas fluorescens* LB 300 or *Enterobacter cloacae* are examples of redox reactions involving enzymatic microbial detoxification of harmful metals or metalloids. Microbes secrete inorganic metabolic products such as sulphide, carbonate or phosphate ions in their respiratory metabolism and

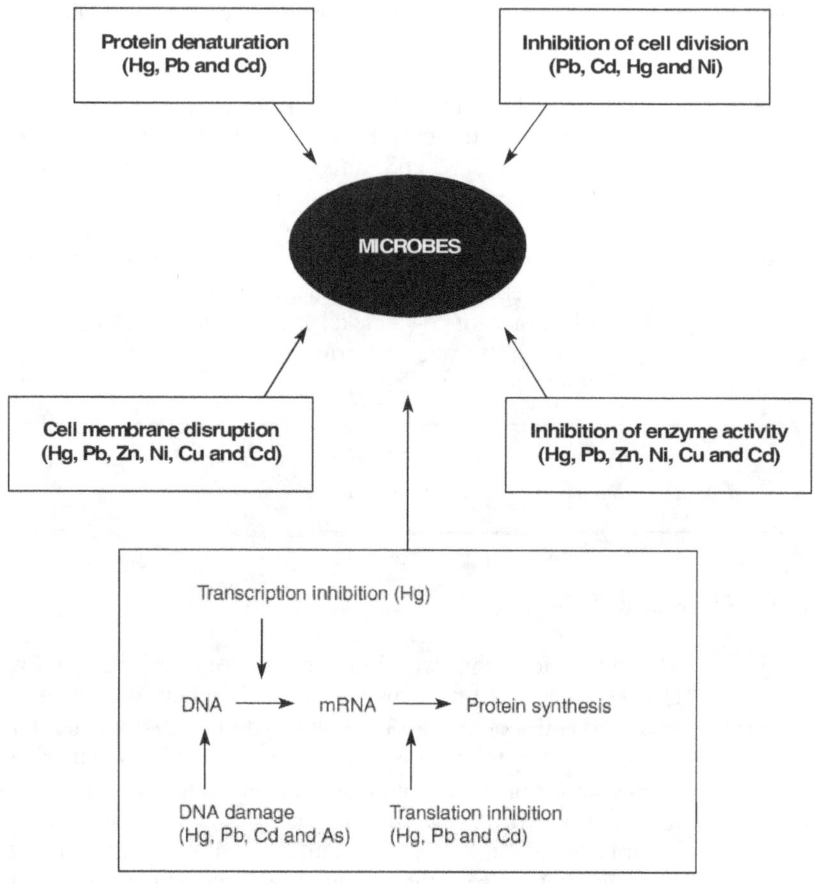

Figure 7.1 Heavy metal toxicity mechanisms in microbes

precipitate toxic metal ions as a form of non-enzymatic detoxification along with these products. In nature, microbes immobilize metals through cellular sequestration and accumulation, or through extracellular precipitation.

Prokaryotes methylate metal and metalloid compounds producing corresponding volatile metal derivatives.

Conventional Method of Heavy Metal Removal from Soil and Water

Various methods are available for the removal and management of heavy metals, which involve technical inputs (Table 7.3). Current methods used for treating soils contaminated with toxic metal are

1. *Land filling* The excavation, transport and deposition of contaminated soil in a permitted hazardous wasteland

2. *Fixation* The chemical processing of soil to immobilize the metals, usually followed by treatment of the soil surface to eliminate penetration by water

3. *Leaching* Using acid solutions as proprietary leaching agents to desorb and leach metals from soil followed by the return of clean soil residue to site (Krishnamurthi, 2000).

Approaches to reduce heavy metal contamination in water include:

1. precipitation or flocculation, followed by sedimentation and disposal of the resulting sludge

2. ion exchange

3. reverse osmosis

4. microfiltration

5. electrodialysis

6. evaporation

Apart from these, other methods like foam floatation, liquid membrane technique, solvent extraction and crystallization can also be employed.

Bioremediation Techniques

Metal-bearing liquid industrial wastes range from high-volume, dilute waters from mine drainages, to aggressive concentrated or multi-component waters from industrial process. Simple metal cations can be removed by biosorption or bioaccumulative processes or biomineralization (Macaskie *et al.*, 2004). The goal of microbial remediation of heavy-metal-contaminated soils and sediments are

Table 7.3 Comparison of conventional and bioremedial metal clean-up strategies

Strategy	Methods	Disadvantage	Remarks
Conventional			
Evaporation	Single/multistage or vapour compression evaporator	Scaling or fouling	High/commercial
Distillation	Packed column with heating and concentration device	Scaling or fouling	Medium/commercial
Solvent extraction	Standard process	Required for the processing	Moderately high/commercial
Adsorption	Batch or continuous adsorption beds	Limited to low concentration	Medium/commercial
Ion exchange	Synthetic product	Requires pretreatment	High/commercial
Membrane process	Standard manufacture units	Separation is imperfect	Medium/commercial
Electrochemical process	DC power and plating apparatus	Impurity upsets the process	Medium/commercial
Starch xanthate process	Synthetic process	Preparation is tedious	Medium/experimental
Bioremediation			
Bioaccumulation	Live microbes, ideal for genetic manipulations	Emerging technology	Lab level
Biosorption	Live or dead microorganism	Emerging technology	Low cost/ commercial
Phytoremediation	Live or dead plant biomass	Emerging technology	Low cost/*Ex-situ* remediation
Plant–microbe interaction	Plant and microorganisms	Emerging technology	Low cost/*Ex-situ* remediation

> *Simple metal cations can be removed by biosorption or bioaccumulative processes or biomineralization.*

to immobilize the metal *in situ* to reduce metal bioavailability and mobility or to remove the metal from the soil. Pazirandeh *et al.* (1997) described the metabolic pathways resulting in bioprecipitation of heavy metals or their biotransformation. Low cost and higher efficiency at low metal concentrations make biotechnological processes very attractive in comparison to physico-chemical methods for heavy metal removal. Metal remediation strategies using microorganisms can minimize the bioavailability and biotoxicity of heavy metals. Biostimulation, stimulation of viable native microbial population, bioaugmentation, artificial introduction of viable population, bioaccumulation, use of live cells, biosorption and use of dead microbial biomass are the cost-effective innovative bioremediation technologies. Biological approach for metal detoxification affords the potential for selective removal of toxic metals and operation flexibility and easy adaptability for *in situ* and *ex situ* application in a range of bioreactor configurations. In the past few decades, new metal treatment and recovery techniques based on biosorption have been explored using both dead and living microbial biomass with remarkable efficiency. Prokaryotic and eukaryotic microbes are capable of accumulating metals by binding them as cations to the cell surface in a passive process.

> *Metal remediation strategies using microorganisms can minimize the bioavailability and biotoxicity of heavy metals.*

Microbes for Metal Remediation

The mechanisms by which metal ions bind to the cell surface include electrostatic interactions, van der Waals forces, covalent bonding, redox interactions, and extracellular precipitation, or a combination of these processes. The negatively charged groups (carboxyl, hydroxyl, and phosphoryl) of the bacterial cell wall adsorb metal cations, which are then retained by mineral nucleation. Biosorption studies of U, Zn, Pb, Cd, Ni, Cu, Hg, Th, Zn, Cs, Au, Ag, Sn, and Mn, showed that the extent of sorption varies markedly with the metal as well as with the microorganisms (Table 7.4).

Prokaryotic and eukaryotic microbes are capable of accumulating metals by binding them as cations to the cell surface in a passive process.

Table 7.4 Example of microorganisms that take up heavy metals

Microorganisms	Elements	Uptake (% dry wt)
Zooglea spp.	Co	25
	Ni	13
Citrobacter spp.	Cd	170
	U	900
Bacillus spp.	Cu	15
	Zn	14
Chlorella vulgaris	Au	10
Rhizopus arrhizus	P	10
	Ag	54
	Hg	58
Aspergillus niger	Th	19

Surfactants such as rhamnolipids produced by *P. aeruginosa* show specificity for certain metals such as Cd and Pb. Higher molecular weight ($\sim 10^6$) bioemulsifiers such emulsan, can also aid in metal removal. Studies of Sand *et al.* (1992) revealed that *Thiobacillus ferrooxidans* and *Leptospirillum ferrooxidans* are capable of oxidizing iron and sulphur. Joerger *et al.* (2001) reported that the metal-accumulating bacterium *Pseudomonas stutzeri* AG 259 is capable of producing silver-based single crystals which can reduce the toxicity of metals.

The mechanisms by which metal ions bind to the cell surface include electrostatic interactions, van der Waals forces, covalent bonding, redox interactions, and extracellular precipitation, or a combination of these processes.

Mechanisms of Metal Tolerance

Organisms respond to heavy metal stress using different defence systems (Figure 7.2) such as exclusion, compartmentalization, formation of complexes and synthesis of binding proteins like metallothioneins (MTs) and phytochelatins (PCs).

Ochari (1997) has divided general toxicity mechanism for metal ions into three categories,

i. blocking the essential biological functional groups of biomolecules especially proteins and enzymes

ii. displacing the essential metal ion in biomolecules

iii. modifying the active conformation of biomolecules resulting in the loss of specific activity

Microorganisms can affect heavy metal concentrations in the environment because they exhibit a strong ability for metal removal from solution; this can be achieved either through enzymatic or non-enzymatic mechanisms. Avoidance, restriction of metal entry into the cell, either by reduced uptake/active efflux or by the formation of complexes outside the cell and sequestration, reduction of free ions in the cytosol either by synthesis of ligands to achieve intracellular chelation or by compartmentalization are the two major strategies of organisms to protect themselves against heavy metal toxicity (Table 7.5).

Table 7.5 Mechanism of metal tolerance in microorganisms

Metal	Tolerance mechanism
AsO_2^-, AsO_4^{3-} and Sb^{3+}	Anion efflux (ATPase)
Cd^{2+} and Zn^{2+}	Efflux (ATPase)
Hg^{2+}	Reduction
Co^{2+} and Ni^{2+}	Efflux
CrO_4^{2-}	Decreased uptake
CrO^{2-}	Decreased uptake
Cu^{2+}	DNA damage

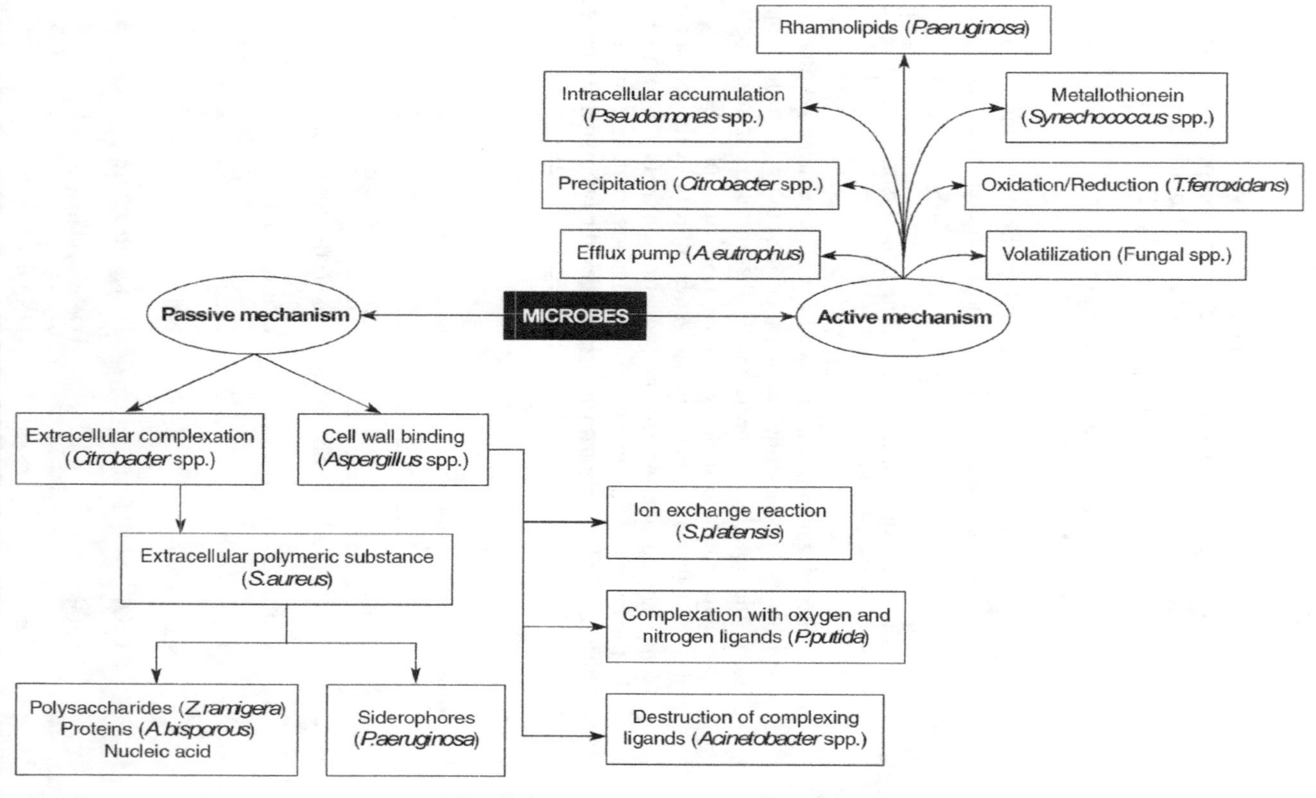

Figure 7.2 Mechanism of microbial metal tolerance

> *Microorganisms can affect heavy metal concentrations in the environment because they exhibit a strong ability for metal removal from solution.*

Metallothioneins

Metallothioneins, discovered about 45 years ago, play a central role in heavy metal metabolism and in the management of various forms of stress. Metallothioneins (MTs) are low molecular weight (6–7 kDa), cystine-rich proteins, divided into three different classes on the basis of their cystine content and structure. The Cys-Cys, Cys-X-Cys and Cys-X-X-Cys, motifs (in which X denotes any amino acid) are characteristic and invariant for metallothionein. Studies of Lichtlen and Schaffner (2001) show that metal regulatory transcriptional factor-1 (MTF-1) was essential for basal and heavy-metal-induced transcription of the stress-responsive metallothionein-I and metallothionein-II. Metallothionein-like proteins have been isolated from *Cyanobacterium, Synecococcus* spp. as well as *Escherichia coli* and *Pseudomonas putida*. Plasmid-encoded energy-dependent metal efflux systems involving ATPases and chemiosmotic ion/proton pumps are associated to Ar, Cr and Cd resistance in *Staphylococcus aureus, Bacillus subtilis, Listeria* spp., *E. coli, A. eutrophus, P. putida*, some cyanobacteria, fungi and algae. *E. coli* use a two-component membrane-bound ATPase complex, ArsA and ArsB responsible for the active efflux of arsenite from the cell. In contrast gram-positive bacteria lack the ArsA protein.

Genetics of Metal Resistance

Recent environment pollution with anthropogenic sources of metals has increased the need for research concerning microbial metal resistance as well as remediation. Large plasmids (165–250 kb), encoded specific metal conferring resistance to a variety of metals including Ag^+, AsO_2^-, AsO_4^{3-}, Cd^{2+}, Co^{2+}, CrO_4^{2-}, Cu^{2+}, Hg^{2+}, Ni^{2+}, Pb^{2+}, Sb^{3+}, TeO_3^{2-}, Tl^+, and Zn^{2+}. Metal resistance systems have been encoded by chromosomal genes in some organisms like, *Bacillus* spp. (Hg-resistance) and *E. coli.* (Ar-efflux). Roane and Pepper (2000) categorized the mechanism of metal resistance as

a. general and do not require metal stress

b. dependent on a specific metal for activation

c. general and are activated by metal stress. The plasmid-borne cadA gene encodes a cadmium-specific ATPase in several **bacterial genera, including** *Staphylococcus, Pseudomonas, Bacillus* and *E. coli.* Schmid and Schegel (1999) reported that *czc* and *ncc* operons are responsible for cadmium resistance in *Alcaligenes eutrophus* CH34 (now renamed as *Ralstonia eutropha*) and in *A. xylosoxidans,* respectively. The genetic determinants for heavy metal resistance in *A. eutrophus* CH34 are present in two plasmids, pMOL 28 and pMOL 30. Nickel resistance is encoded by pMOL 28 (163 kb) which occur by an efflux pathway via cation proton–antiporter chemiosmotic system (Mergeay, 1991). This removes the toxic metal, which is accumulated by the uptake of essential divalent cations (e.g. Mg^{2+}, Mn^{2+}). Taghavi *et al.* (2001) showed that nickel resistance is inducible and is due to an energy-dependent efflux system driven by chemiosmotic proton–antiporter system.

Genetic Engineering for Metal Remediation

Genetic engineering allows introduction of desired traits into cells to metal clean up techniques, and this approach has already been used to construct cells for the bioremediation of mercury. Krishnasamy and Wilson (2000) constructed an *E. coli* strain that accumulated Ni^{2+} by introducing the *nixA* gene (coding for a nickel transport system) from *Helicobacter pylori* into *E. coli* JM109 that expressed a glutathione S-transferase pea metallothionein fusion protein. The recombinant *E. coli* strain accumulated four times more nickel than the wild type. Bang *et al.* engineered *E. coli* to produce sulphide by heterologous expression of the thiosulphate reductase gene from *Salmonella enteritica,* which enhanced protein synthesis leading to the precipitation of cadmium sulphide. Metal resistant *R. eutropha* isolate, engineered to produce metallothionein accumulated more Cd^{2+} than its wild type counterpart, and offered tobacco plants some protection form Cd^{2+} when inoculated into contaminated soil. Studies of Pazirandeh (1998) showed periplasmic expression of a *Neurospora crassa* metallothionein in *E. coli-* generated cells that were superior to bacteria with cytoplasmic metallothionein location in terms of metal ion adsorption. Kotriba *et al.* engineered metal-binding peptides that contain either histidines $(GHHPHG)_2$ (HP) or cystines (GCGCPCGCG) (CP) into LamB and expressed on the surface of *E. coli.* Surface display of CP and HP increased the bioaccumulation four- and two-fold, respectively.

> *Genetic engineering allows introduction of desired traits into cells to metal clean up techniques.*

Samuelson *et al.* (2000) reported that recombinant *Staphylococcus xylosus* and *S. carnosus,* had gained Ni^{2+}- and Cd^{2+}-binding capacity and suggested that they could be used in bioremediation of heavy metals. They evaluated surface display systems for expression of two different polyhistidyl peptides, His_3-Glu-His_3 and His_6, which had good metal-binding activity. The nickel-resistant bacteria from anthropogenically polluted biotypes and naturally nickel-rich soils have been hybridized with various probes carrying *cnr* from plasmid pMOL28 (*A. eutrophus* CH34), *ncc* from pTOM8 (*A. xylosoxidans* 31A) *nre* from pTOM8 (*A. xylosoxidans* 31A) and *nre* from *Klebsiella oxytoca* 15788. With respect to their hybridization signals, the nickel resistant determinants of the strains could be assigned to *cnr/ncc* type, *cnr/ncc-nre* type *Klebsiella oxytoca* type and others showing no hybridization. Bacterial communities in soil amended for many years with sewage sludge that contained heavy metals were assessed using molecular tools such as rRNA analysis, FISH (Fluoresence *in situ* hybridization), cloning and sequencing.

Mechanisms of Metal Remediation

1. *Biosorption*

Biosorption which is the utilization of inexpensive dead or live microbial biomass to sequester metals from industrial effluents has gained importance in recent years due to their good performance, low cost, specificity, minimum sludge generation and amenability for repeated use. Several active groups of cell constituents like acedamido group of chitin, structural polysaccharide of fungi, amine (amino and peptidoglycosides), sulphahydral and carboxyl groups in protein, phosphodiester (teichoic acid), phosphate, hydroxyl in polysaccharides, participate in biosorption. Hernandez *et al.* (1998) isolated three species of bacteria belonging to the family Enterobacteriaceae, which were capable of accumulating nickel and vanadium.

> *The technology involving surface complexation, ion exchange and microprecipitation is a potential alternative to current metal-treating processes that extend the duration of exposure of the clean-up crews.*

Biosorption for metal removal has been studied extensively using various species of live and inactivated biomass of bacteria, algae, fungi or yeast. The technology involving surface complexation, ion exchange and microprecipitation is a potential alternative to current metal-treating processes that extend the duration of exposure of the clean-up crews. Research in the area of heavy-metal removal from waste water and sediments has focused on the development of microbial materials with increased affinity, capacity and selectivity for target metals. Fungal biomass is finding increasing application in current biotechnological procedures for the clean-up of nickel-polluted effluents and for the recovery of metal ions. Rajendran *et al.* (2002) reported *A. niger* mutant M3 capable of absorbing 50% more nickel at 1.7 m*M* concentration than its parent strain. The waste fungal biomass from industrial fermenters may prove to be cost-effective metal-biosorptive agents on a large scale. Fourest and Roux (1992) studied the use of mycelia of *Mucor nicheri, Aspergillus niger* and *Penicillium chrysogenum* from fermenters in the removal of nickel, zinc, cadmium and lead by bio-adsorption. Gadd (1992) reported that chemical modification of biomass might create derivatives with altered metal-binding abilities and affinities. Pseudomonads have been shown to be efficient in heavy-metal bioaccumulation from polluted effluents in the immobilized state.

ii. *Immobilization for metal remediation*

Biosorption technique for metal removal and accumulation can be carried out even with a low degree of understanding of the metal binding mechanisms. However, a better understanding of the technique will be useful for effective and optimized application of the method. Free cells can provide valuable information in laboratory experimentation but have limitations in industrial applications. The use of freely suspended microbial biomass suffers from disadvantages like small particle size, low mechanical strength and difficulty in sequestering biomass and effluent. The immobilized bioadsorbent biomass packed in sorption columns which are effective for continuous removal of heavy metals can operate on cycles consisting of loading, regeneration and rinsing. Ibanez and Umetsu (2002) have demonstrated the ability of protonated alginate beads in the removal of chromium, copper, zinc, nickel and cobalt ions from dilute aqueous solutions. Biomass immobilized in a range of inert materials like silica, polyacrylamide, polymethane and polysulphone has been used in a variety of bioreactor configurations, including rotating biological contractors, fixed reactors, trickle filters, fluidized beds and air lift bioreactors. Karna *et al.* (1999) immobilized *Phormidium valderianum* BDU 30501 in polyvinyl foam and used for the removal of Ca^{2+}, Co^{2+}, Cu^{2+} and Ni^{2+}. Asthana *et al.* reported that Ca-alginate immobilizing agent is very effective in nickel biosorption and used inorganic salts like NaCl and Ca $(NO3)_2$ for desorption of nickel from immobilized microbial biomass. HCl and EDTA are the most efficient desorption agents for nickel removal.

> *The immobilized bioadsorbent biomass packed in sorption columns which are effective for continuous removal of heavy metals can operate on cycles consisting of loading, regeneration and rinsing.*

III. *Plant–microbe interaction in metal remediation*

Heavy metal toxicity to plants can be reduced by the use of plant growth-promoting bacteria, free-living soil bacteria that exert some beneficial effects on plant development when they are applied to seeds or incorporated into seeds. Plants can accelerate bioremediation in surface soils by their ability to stimulate soil microorganisms through the release of nutrients from the soil and transport of oxygen to the rhizosphere. Plant growth-promoting bacteria that contain ACC deaminase may act to ensure that the ethylene level does not impair root growth and facilitate the formation of larger roots, which enhance seedling survival. Burd *et al.* (1998) reported that a metal-resistant soil bacterium *Kluyvera ascorbata* SUD 165 promoted the growth of canola (*Brassica campestris*) in the presence of high concentrations of nickel and the increase in growth rate is due to the ability of the bacterium to lower the level of ethylene stress in the seedlings. They further isolated siderophore over producing mutants, *K. ascorbata* 165/26 which was able to provide sufficient ions for the growth of Indian mustard

(*Brassica juncea*) and tomato (*Lycopersicon esculentum*) in the presence of high concentrations of Ni, Pb, and Zn. Studies of Burd *et al.* (2000) suggested that the best method to prevent plants from becoming chlorotic in the presence of high levels of heavy metals was to provide them with a siderophore-producing bacterium, *Kluyvera ascorbata* SUD 165. Mycorrhizal fungi also reduce metal toxicity to their host plant by binding heavy metals to their cell wall or surrounding polysaccharides. Binding sites within cells of ectomycorrhizal fungi are metallothionein-like proteins, which have high cystine content and perhaps polyphosphate granules.

> *Plants can accelerate bioremediation in surface soils by their ability to stimulate soil microorganisms through the release of nutrients from the soil and transport of oxygen to the rhizosphere.*

Field Applications

The most important biotechnological application of metal–microbe interaction is in bioleaching, bioremediation of polluted sites and mineralization of polluting organic matter. Uranium is extracted from uranite ore through bioleaching by indirect action with *T. ferrooxidans*. In this case, ferric iron in acidic solution, which the organism generate when oxidizing pyrites (FeS_2), oxidizes insoluble UO to soluble UO_2^{2+}. The extraction of metals such as Co, Mo, Ni, Pb, and Zn from sulphide ores by bioleaching is technically feasible. Gencor (South Africa) is currently developing the BioNIC process to recover nickel from low-grade sulphide ores. Signet Technology Inc. and Signet Engineering Pvt. Ltd. (Western Australia) are currently developing a process to recover cobalt from pyrite ore from the kares cobalt project in Uganda. Various microbially reducible metals, especially ferric iron in complexed form to keep it soluble at circumneutral pH, can be used as terminal electron acceptors in *in situ* anaerobic bioremediation of sites polluted with toxic organics. It is possible to bioremediate *in situ* sites polluted with chromate or dichromate [Cr (VI)] by stimulating the reduction of the Cr (VI) to Cr (III) by bacteria[25]. Fungi can convert oxidized selenium to volatile methylated selenides for escape into the atmosphere, a removal by volatilization. Bosecker (1993) reported that *Penicillium funiculosum* was able to extract more than 50% Ni and 75% Zn from test solutions containing 100 mg l^{-1} of metal at pH 6.6 and 6.5 respectively.

A potent algal biosorbent Alga Sorb™ developed using a freshwater alga *Chlorella vulgaris* and AMT-BIOCLAIM™ (MRA) developed using *Bacillus* biomass are used to treat waste water and metal recovery respectively.

Future Prospects

Microbes can significantly affect the distribution of metals in the environment, since they have developed means to use them for their benefit. This clearly holds promise for effective, economical and eco-friendly metal bioremediation technology for industrial exploitation and pollution-free environment. A good and efficient metal biosorbent can replace the commercial ion-exchange resins that have been used conventionally for metal removal. However, the basic knowledge of microbial metal bioremediation mechanism is inevitable for the development of commercially viable potent biosorbent. Although several microbial metal bioremediation approaches are established to combat heavy metal pollution from anthropogenic and natural sources, none is yet in widespread use. Innovative, economically feasible and novel biomass regeneration and conversion of the recovered metal into usable form are the best options to attract more usage of biosorbents. It is time to initiate a more comprehensive interdisciplinary approach between biotechnologists and metallurgists to bring lab-scale bioremediation process to large-scale technology that will be acceptable to industrialists.

> *A good and efficient metal biosorbent can replace the commercial ion-exchange resins that have been used conventionally for metal removal.*

PESTICIDE DEGRADATION

Environmental damage due to pesticides is a serious problem. The complex chemistry of pesticides, their persistence in the environment, toxicity to animals and human beings, and bioaccumulation risks make pesticide contamination a critical problem. The tremendous pesticide chemistry makes their detoxification process a difficult task. Physical methods (incineration, entrapment and burial) and chemical methods (solvent extraction, oxidation–reduction and hydrolysis) were developed to control pesticide waste discharges. Pesticides enter into the environment by (i) manufacturing wastes—point discharge, (ii) user wastes and (iii) residue wastes (Figure 7.3). The discharge of pesticide wastes originating from manufacturing plants can be subjected to treatment and control. By-products arising during manufacturing and discarded residues may become part of solid or liquid effluent. Increased persistence in surface soil increases the possibility of pesticides being carried to ground water. Pesticide sorption, degradation and water movement are the three processes which control pesticide transport to the ground water. Pesticides with moderate to long half-lives in the soil, low sorption potential and high water solubility have more changes to reach the ground water.

> *Pesticide sorption, degradation and water movement are the three processes which control pesticide transport to the ground water.*

Current Pesticide Scenario

The first widespread use of insecticide use began at the end of World War II. Since 1945, chemists have developed many kinds of synthetic organic chemicals for use as pesticides. Worldwide about 2.3 million metric tons of these pesticides are used each year on an average of 0.45 kilogram for each person on earth. About 85% of all pesticides are used in developing countries (Joseph, 2005). The contamination related to agricultural inputs has been growing mainly due to the extended use of pesticides. They affect the soil microflora and chemistry. Pesticides even though present in exceedingly low concentration in the surrounding water are taken by various microbes, plants and animals which may accumulate and concentrate them several thousand times. The process of bioaccumulation and biomagnification makes exceedingly low quantities of pesticides or their toxic residues available to the living organisms in highly concentrated state. Safe and economical disposal of pesticides is a current problem of significant magnitude. Continuous use of pesticides creates resistance to their action in pests and leads to further modification and production of new chemical forms.

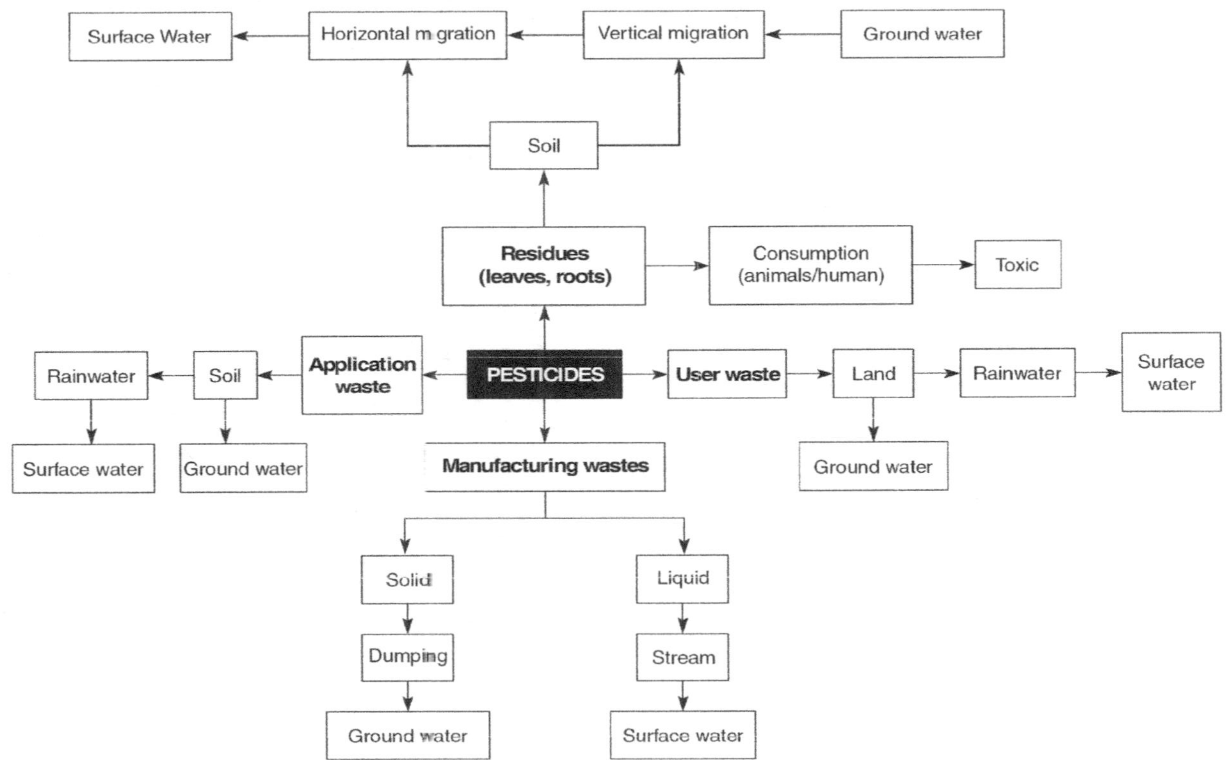

Figure 7.3 Pesticide contamination routes

Pesticide Classification and Disposal

Pesticides can be classified in a number of ways based on their target (insecticides, herbicides, fungicides, rodenticides, algicides, molluscicides, nematicides, etc.), chemical nature (organophosphates, chlorinated hydrocarbons, carbamates, pyrethroids, carbamic acid derivatives, nitrophenols, etc.) and mode of action (stomach poisons, contact poisons, fumigants, etc.). Pesticide disposal through physical, chemical and microbiological means has received serious attention the world over (Figure 7.4). Increased persistence in surface soil increases the possibility of pesticides being carried to ground water. Numerous environmentally compatible pesticide disposal and decontamination techniques have been developed that includes construction of disposal pits, evaporation beds, filtration systems, photo-decomposition, UV-ozonation and microbial processes (Walia and Parmar, 1993). Among the various physical, chemical and biological processes, microbial degradation is the major pathway by which pesticides and related xenobiotics are eliminated from the ecosystem.

Microbes for Pesticide Degradation

Physical, chemical, photochemical and microbial actions with a consortium of microbes are responsible for the degradation of pesticides in the environment (Table 7.6). Biodegradation ability in microbes is achieved through exposure and adaptation or mutation and selection or through rDNA technology. Microbes possessing manipulated genes or enzymes with specific degradative abilities may be used for biodegradation. Insertion of pesticide-degradative gene into the host microbes indigenous to the environment will be advantageous since it can tolerate solvent, thermal, saline and other stresses. Low bacterial population, less nutrition (dissolved organic carbon) and low oxygen levels of vadose zone and aquifer are the limiting factors for biodegradation of pesticides. Fungi and bacteria can degrade majority of pesticides. Microbial consortium can transform parent pesticide molecules to less complex products (Figure 7.4). The extreme toxicity of the pesticide is often lost in the first transformation step. Hydrolysis reduces toxicity and enhances biodegradability changes for the transformed pesticide. Microbial degradation of pesticides are brought mainly by the enzymes located on endoplasmic reticulum, which is a network of lipoprotein membranes within a cell (Asthana and Asthana, 2001). Enzymes such as esterases, acylamidases, phosphoesterases can also be used for pesticide degradation. Biodegradation of pesticides through dehalogenation, dehydrohalogenation, oxidation, reduction, hydrolysis, isomerism and rearrangements, etc., is influenced by the nature of microbial strains, diverse structural features of pesticide molecule and other soil variables.

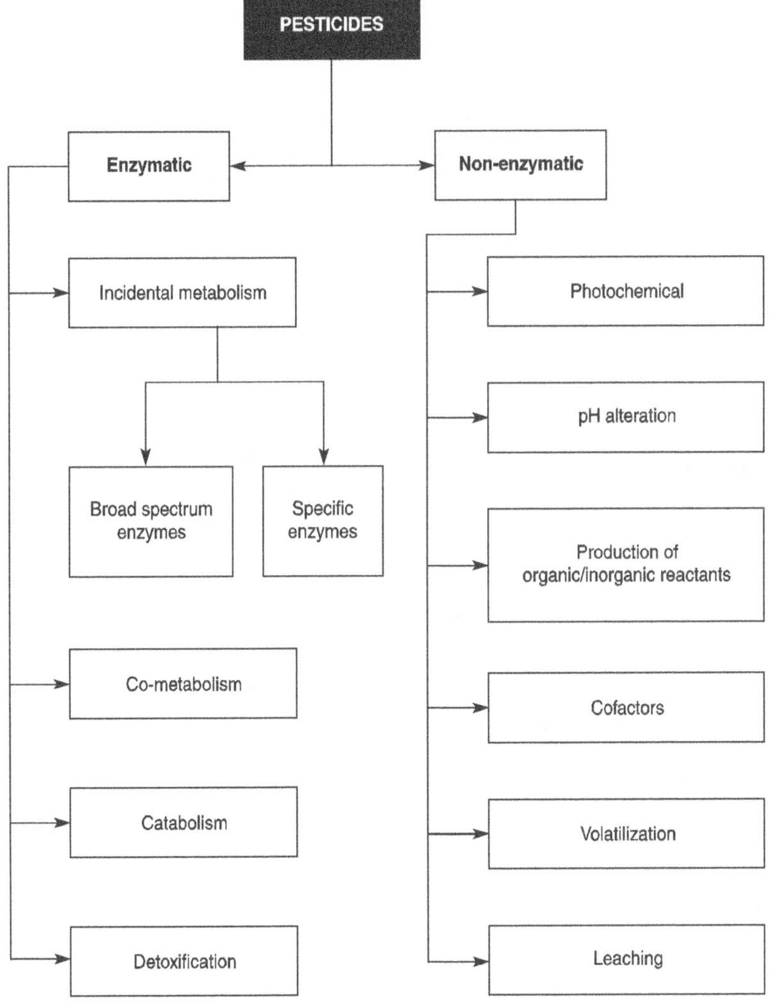

Figure 7.4 Pesticide degradation methods

Table 7.6 Microbes involved in pesticide and herbicide degradation

Insecticide	Organism	Mechanism	Enzyme
Coumaphos	*Flavobacterium* spp.	Hydrolysis	Parathion hydrolase
Parathion	*Pseudomonas* spp.	-	Parathion hydrolase
Diazinon	-	-	Cell-free enzyme extract
Propanil	*Fusarium solani*	-	Aclamidase
Malathion	*Trichoderma viridae*	Hydrolysis	-
4-methyl phenol	*Geotrichum candidum*	-	Laccase
DDT and Lindane	*Phanerochaete chrysosporium*		-
3,4 Dichloroaniline	*Phanerochaete chrysosporium*	Mineralization	-
Dieldrin	*Chrysosporium lingorum*	Mineralization	-
Phenanthrene	*Trametes versicolor*	Mineralization	-
	Sternum hirsutum		
2,4 DES	*Pseudomonas putida*	Cleavage of O-S and C-O-S linkages	Alkyl phosphatase
Chlorobenzonate herbicides	*P. putida, C. tropicalis*	Degradation	-
	F. flocciferum A. niger	Degradation	-
	A. japonicum Aureobasidium pullulans	Degradation	-

(Contd.)

Table 7.6   (Continued)

Insecticide	Organism	Mechanism	Enzyme
Methoxychlor	*B. acinetobacter, Rhodococcus*	Dechlorination Demethylation	-
Aldrin, Endrin	*Clostridium bifermentans C. glycolium*	Dehalogenation	Enzymes
Dithioate	*E. coli*	-	Cell-free enzyme extract
Chlorpropham	*Bacillus sphericus*	-	Amidase
Linuron Monolonuron	*Penicillium* sp.	-	Acylamidase
Karsil	*Fusarium solani, F. oxysporum*	-	Acylamidase
3,4-D	*Arthrobacter* sp.	-	Hydroxylase
MCPA	*Flavobacterium perejunum*	-	Cell-free extract

Physical, chemical, photochemical and microbial actions with a consortium of microbes are responsible for the degradation of pesticides in the environment.

Pesticide Degradation Mechanisms

Detoxification of persistent pesticides can be achieved by the enhancement of co-metabolism induced by the introduction of appropriate primary growth substrates in the medium (Figure 7.5). This approach is particularly suitable for poorly degradable contaminants and for toxicant present at concentration too low to induce degradative activity. When local environmental conditions are unable to develop degradative microbial population at the contaminated site, it is advantageous to inoculate the sub-surface with established and well-tried degradative microorganisms. Enhanced degradation of pentachlorophenol in topsoil was achieved in field conditions following inoculation of the pesticide with PCP-degrading *Arthrobacter* (Edgehill and Finn, 1983).

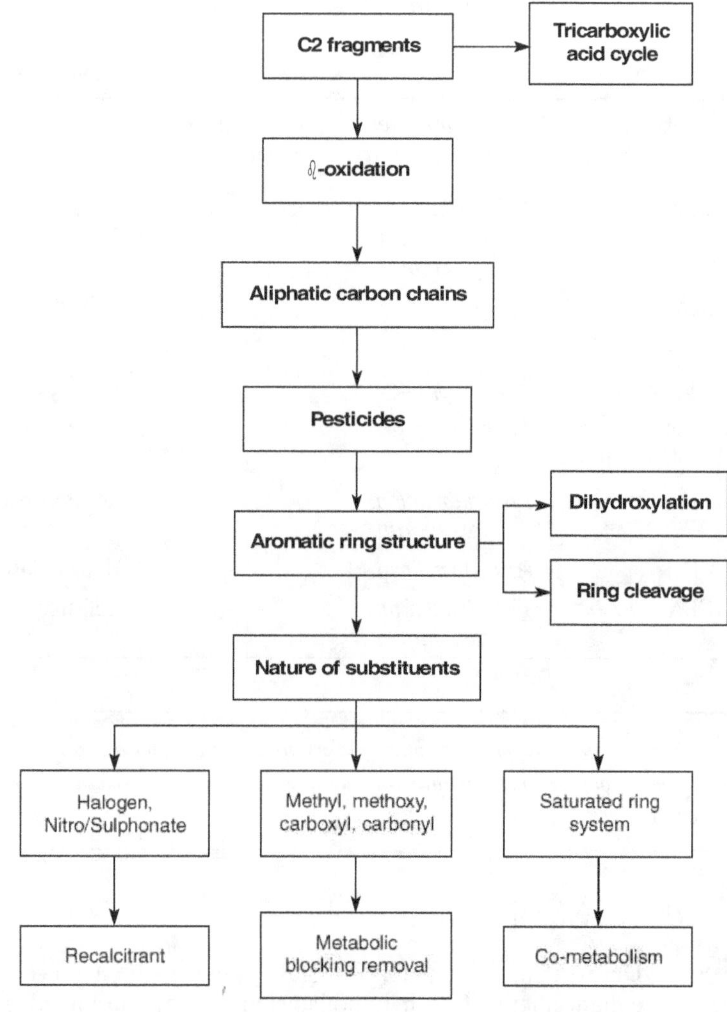

Figure 7.5 Mechanisms of pesticide degradation pathways

Pesticide Biotic Transformation

Microbes such as bacteria, algae fungi, etc. are able to perform diverse biochemical reactions at a much faster rate per unit body weight as compared to higher plants and animals. Diversity in the microbial species composition ensures a wide variety of pesticides to be

> *Microbes such as bacteria, algae, fungi, etc. are able to perform diverse biochemical reactions at a much faster rate per unit body weight as compared to higher plants and animals.*

subjected to the degradation activity. Each microbe performs its own specific task to bring about the complete decomposition of pesticides. Biotic transformation involves biochemical reaction brought about by the enzymatic machinery of living organisms. Pesticides are lipophilic in nature, absorbed by the living system and cannot be excreted. They have to be converted into hydrophilic compounds by the action of enzymes. In certain cases the products of enzyme action will be more toxic than the original compounds. Biochemical reactions involved in transformation can be categorized into degradative and conjugative reactions. Degradative reaction involves oxidation, reduction, hydrolysis, cleavage of important bonds, etc. The reactions result in displacement of functional groups of the pesticide molecules such as $-OH$, $-COOH$, and $-NH_2$. Conjugation reaction involves a synthetic reaction in which the pesticide molecules, carrying groups acquired during the earlier reactions, are conjugated to other molecules within the living system. They form highly ionized water-soluble substances that can be quickly excreted from the living system.

> *Pesticides are lipophilic in nature, absorbed by the living system and cannot be excreted.*

Enzyme Detoxification

Immobilized pesticide-degrading enzymes can play a vital role in detoxification and disposal of a variety of pesticide wastes. The enzymes that can function without the involvement of cofactors and coenzymes have a great potential for degrading pesticides by hydrolytic, oxidative or other simple enzymatic reactions. Organophosphorous, carbamate, acylanilide, phenyl urea and pheno-oxyacetate pesticides can be detoxified by the enzymes such as hydrolase and esterase.

> *Immobilized pesticide-degrading enzymes can play a vital role in detoxification and disposal of a variety of pesticide wastes.*

Plasmids in Pesticide Degradation

Recent developments in plasmid technology and cloning technique have improved the pesticide degradation abilities of microbes. Bacterial plasmids have resulted in the degradation of more persistent and toxic herbicides 2, 4-D, 2, 4, 5-T, MCPA and many naturally occurring aromatic and aliphatic compounds. The use of an undefined mixture of cells isolated from sites contaminated with 2,4,5-T resulted in the selection of strains capable of utilizing this herbicide as the sole carbon and energy source. The DDT-degrading strain of *Pseudomonas* conjugated with the strain carrying naphthalene-metabolizing plasmid, resulted in strains that were capable of metabolizing DDT.

Factors Affecting Pesticide Degradation

Pesticide degradation in soil surface is affected by: soil type, surrounding conditions, population of microbes and their metabolic capabilities, pH, moisture, organic contents, temperature, aeration, and cation exchange capacity. Several other factors such as structure and solubility, low solubility of highly chlorinated compounds, halogen substitution, polar groups such as $-OH$, $-COOH$ and $-NH_2$, and heavy metal concentration, etc. also affect biodegradation of pesticides. The resonance structure of benzene nucleus in aromatic pesticides is responsible for its inertness. Microbes can degrade them only when the energy barrier gets lowered. Microbial enzyme catalyses ring fission and subsequent reactions release the energy. In aerobic systems, microbes make use of oxygen to hydroxylate the benzene ring and facilitate its cleavage whereas in anaerobic system, resonance caused by unsaturated bond is removed by reduction.

> *In aerobic systems, microbes make use of oxygen to hydroxylate the benzene ring and facilitate its cleavage whereas in anaerobic system, resonance caused by unsaturated bond is removed by reduction.*

SYNTHETIC POLYMERS

Polymers are macromolecules, that is, molecules containing many atoms that are of very large size and high molecular weight (10 Kda to over 100 Kda). They are made up of a small number of repeating groups. Synthetic polymers are made up of a great number of simple units (monomers) joined together in a regular fashion. Most of the biodegradable polymers contain functional groups that are subject to attack by microbial enzymes. Polymers contain functional groups that are subject to attack by microbial enzymes. Polymers contain ester or amide groups that can be hydrolysed to a linear chain that can be oxidatively cleaved. Microbes are unable to ingest high polymers through their cell wells. They cannot ingest a molecule that has a molecular weight greater than about 500 Da. Polyester-degrading microbes secrete esterases, enzymes that catalyse the hydrolysis of ester groups, which break the polymer into smaller fragments that can then be ingested. Commercial synthetic polymers undergo biodegradation in the presence of molecular oxygen.

> *Polyester-degrading microbes secrete esterases, enzymes that catalyse the hydrolysis of ester groups, which break the polymer into smaller fragments that can then be ingested.*

Biodegradation of Polymers

Polymer degradation in the biosphere is limited because of its extreme stability. Approximately 140 million tonnes of synthetic polymers are produced worldwide every year. Synthetic polymers such as waste plastics and water-soluble synthetic polymers in waste water have been recognized are major environmental pollutants. The biodegradability depends on the molecular weight, molecular form and crystallinity. It decreases with increase in molecular weight, while monomers, dimers and repeating units degrade easily. Two categories of enzymes are involved in the degradation process, namely extracellular and intracellular depolymerases. Exoenzymes of microbes first break down the complex polymers giving short chains that are small enough to permeate through the cell walls to be utilized as carbon and energy sources. The process is called depolymerization. When the end products are CO_2, CH_4, or H_2O the process is called mineralization (Figure 7.6).

> *Synthetic polymers such as waste plastics and water-soluble synthetic polymers in waste water have been recognized as major environmental pollutants.*

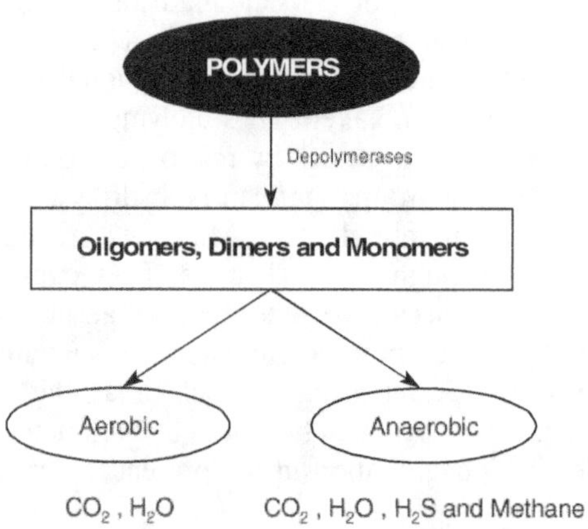

Figure 7.6   Reaction pathways during biodegradation of polymers (Premraj & Doble, 2005)

Plastic degradation   Plastics are polymers of high molecular weight and are usually hydrophobic. Synthetic plastics are not degradable and are not compatible with natural biogeochemical cycling of elements. They cannot be composted with other organic solid wastes. Disposed plastic materials end up in landfills and they are resistant to anaerobic decomposition. Incineration of plastics release toxic compounds into the atmosphere. Synthetic plastics are made from petrochemicals that are not renewable. Accumulated plastic materials in the soil disintegrate into fine particles or powder (the molecular size and structure of the polymer do not change). The hydrophobic nature of plastic material alters the physical and chemical nature of the soils by reducing the water-holding capacity, which in turn would reduce the microbial population in the soil.

ASTM (American Society for Testing and Materials) and ISO (International Organization for Standardization) have devised certain protocols to produce plastics. Plastics can be degraded photochemically (break up of chemical bonds in the polymer by ultraviolet light energy), chemically, (breaking of chemical bonds by oxidation or hydrolysis in water) and biochemically (chemical bond breaking by enzymes). A remarkable characteristic of microbial polyesters is their biodegradability in the environment. The films and fibres of microbial polyesters can be degraded in soil sludge, freshwater pond, compost and sea water (Mergaert, 1992). Chemical bonds of polymers are broken down into oligopolymers by irradiation energy or oxygen. They are amendable to mineralization or co-metabolization by microbes. These smaller molecules may be recalcitrant and bring about problems such as toxicity, increased mobility and solubility, and biomagnification.

Alternative to synthetic plastics   Since plastic has become a part and parcel of modern life, it has become imperative to develop biodegradable materials. Biodegradable polymers are defined as those whose chemical bonds are broken by microbial enzymes, which either mineralize the polymer or reduce it to intermediates that can be metabolized by microorganisms and plants (Tan, 2004). They are biocompatible (do not induce an adverse reaction from cells and tissues in contact with them, such as inflammation, tumour development, or tissue rejection). A biodegradable polymer must have the following properties:

    a.   it should degrade readily and completely after disposal, and not leave toxic or hazardous intermediates in the natural environment

    b.   it should be synthesized from renewable resources

    c.   the new material should have properties and application similar to synthetic plastics

Polyhydroxyalkanoates (PHA) from microbial origin are considered good substitutes for plastics and elastomers, since in properties they are similar to petrochemical plastics, yet are truly biodegradable (Revite & Mavinkurve, 2001). Though several bacteria accumulating PHA have been identified, the commercial production of these polyesters has been limited to only one product Biopol, because of the high price as compared to the petrochemical plastics.

Polyhydroxyalkanoates (PHA)   The desirable characteristics of PHAs are

    i.   it is a thermoplastic and has a wide range of chemical and material properties

    ii.   it is biodegradable, non-toxic and biocompatible

iii. it can be synthesized by bacteria and plants and there is no need for additional chemical processes for polymerization

iv. it can be produced from renewable resources. PHA is polyester produced naturally by many bacteria, usually as a carbon and energy storage material (Figure 7.7). The molecular weight ranges from 2×10^5 to 3×10^6. It is composed of repeating units of 3-hydroxyl alkanoic acids, each of which carries an aliphatic side-chain (R).

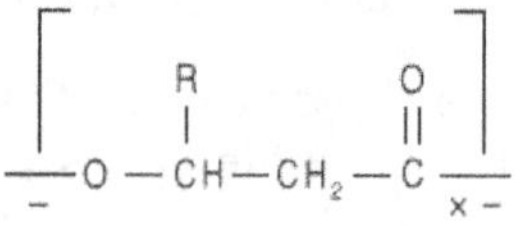

$$\left[\ -O-CH-\underset{\substack{|\\R}}{}CH_2-\overset{\substack{O\\||}}{C}\ \right]_{x}$$

Figure 7.7   General structure of a repeating unit of poly-3-hydroxyalkanoate. R = —$(CH_2)_n$ —CH_3 where n = 0 to 10 or more x = 4,000 to 20,000

There are two main classes of PHA: the short-chain-length (scl) variety in which monomers are 4 to 5 carbons in length and the medium-chain-length (mcl) PHA in which monomers are 6 to 14 carbons length, e.g. for Scl-PHA is poly (3-hydroxy butyrate) PHB and the poly (3-hydroxybutyrate-Co-hydroxyacetate) PHBV for mcl-PHA are 3-hydroxy hexapote, 3-hydroxy octanoate, and 3-hydroxydecanoate monomers. The Scl-PHA are accumulated in *Ralstonia eutropha, Azotobacter vinelandii, Alcaligenes latus* and *Burkholderia cerpacia.* mcl and Scl-PHA synthesizing bacteria have different sets of PHA-synthesizing enzymes. There are three principal enzymes involved in PHA synthesizers, viz β-ketothiolase, acetoacetyl CoA reductase and PHA synthase. The mcl-PHA monomers are directly derived from the carbon substrate (alkanoic acids, alkanes, alkanols) or from intermediates arising from the beta-oxidation of long-chain carbon compounds. Fluorescent pseudomonad group of bacteria is known to produce mcl-PHA, e.g. *Pseudomonas oleovorans* and *P. putida*. The mcl-PHA is generally amorphous and elastomeric. Physical properties of the polymer such as thermal properties, molecular weight, viscosity crystallinity and mechanical properties, vary with the composition of PHA. PHA exhibits a strong absorbance at 1720 and 1180 cm^{-1} in the infrared spectrum, generally representing the ester linkage. Characterization of the polymer can be done by GC analysis followed by HNMR and CNMR spectroscopy.

PHA Degradation

1. *Intracellular degradation* Bacteria accumulating PHA possess an enzyme system, which appears to be rather complex, consisting of several components, mobilizing PHA intracellularly under certain conditions. The predominant products of PHB hydrolysis are 3-HB and its dimeric esters. Two proteins, H and I exhibit a high degree of homology to the HPr protein and enzyme 1 of the phosphoenolpyruvate–carbohydrate phosphotransferase system.

2. *Extracellular degradation* Synthesis of extracellular PHA hydrolases in bacteria and fungi can be demonstrated by a clearance zone around the colony on mineral salt agar medium containing PHA as the sole carbon source. PHB-degradation occurs in two steps: the hydrolysis of PHB from the hydroxyl terminus to release 3HB dimers followed by their cleavage into the monomers, as demonstrated in *A. faecalis*. Anaerobic degradation of PHA has been observed in *Iiyobacter delafieldii, I. polytrophus* and *Clostridium homopropionicum* yielding acetate, butyrate, methane and molecular hydrogen.

Plasticizers

Plasticizers are the compounds that impart plasticity and mobility to synthetic polymers and plastics. Phthalate esters like di-2-elthylhexyl phthalate and di-*n*-butyl-phthalate are the important plasticizers. The plasticizers migrate into the product from the bugas and containers used as packing materials from where they are released into the environment. Phthalate ester plasticizers are relativity stable, recalcitrant compounds with a low acute toxicity. They are teratogenic, carcinogenic and complexed with fulvic acid component of humic substances in the soil as well as aquatic bodies.

Anaerobic Degradation of Phthalates

Bio (2-ethylhexyl) phthalate accounts for about 23% of phthalate production and it is a suspected carcinogen. Phthalate esters (Figure 7.8) are degraded slowly under anaerobic conditions with less degradation for longer side-chains: 100% loss in 7–42 days for methyl, ethyl, butyl and butylbenzylphthalates, but insignificant loss of octyl and ethylhexylphthalates. The proposed pathway for the degradable phthalate esters is by cleavage of a side chain subsequently, forming *o*-phthalate, the ring cleavage by an unknown pathway (Figure 7.9) Phthalic acids are degraded by decarboxylation through benzoyl CoA and pimelate as in the degradation of phthalates by a denitrifying bacterium. Anaerobic pathways leave hexylethylphthalate untransformed in digested municipal sludge.

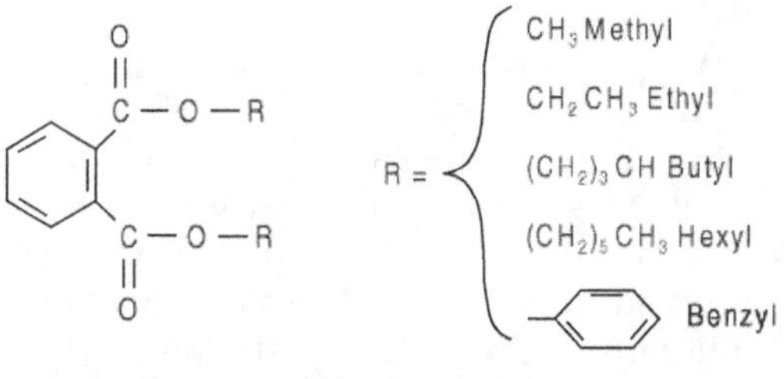

Figure 7.8 Structure of phthalates

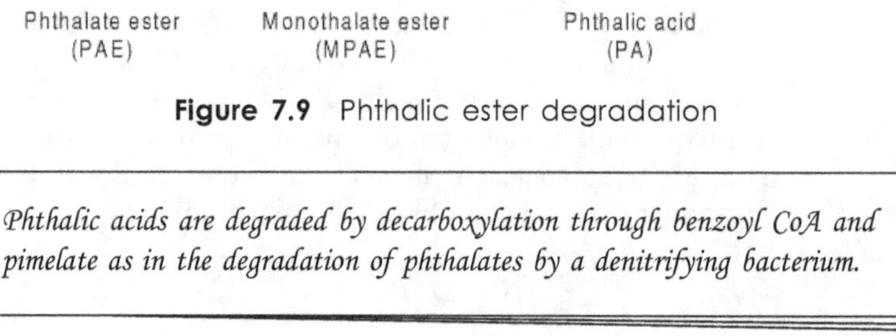

Figure 7.9 Phthalic ester degradation

> *Phthalic acids are degraded by decarboxylation through benzoyl CoA and pimelate as in the degradation of phthalates by a denitrifying bacterium.*

DIOXINS

Dioxins are a group of chemical compounds that occur accidentally as contamination in a number of industrial processes and products. They are highly toxic, and carcinogenic to humans and animals.

They are formed due to the incomplete incineration of waste and the burning of plastics, coal or cigarettes. Dioxins are deposited on plants, soil and water then by entering the food chain. They are also produced during the manufacture of paper. The pearly white colour of the paper is the result of bleaching. Bleaching uses chlorine, which produces dioxin. The toxic effluent from the paper factory ends up in water bodies. They are persistent polychlorinated pollutants in the dioxins. Polychlorinated dibenzo-*para*-dioxins (PCDDS) are present as trace impurities in some manufactured chemicals and industrial wastes. The chemical and environmental stability of PCDDs and their tendency to accumulate in fat have resulted in their detection within many ecosystems. In general, wherever high levels of PCDDs have been detected, the source has been a hazardous waste dump, an industrial discharge, or an application of PCDD-contaminated herbicide. There are 75 PCDD isomers; some are extremely toxic, while others are believed to be relatively innocuous. The most toxic and most extensively studied PCDD isomer is 2,3,7,8-tetrachlorodibenzo-para-dioxin (2,3,7,8-TCDD). In the United States and elsewhere, accidental contamination of the environment by 2,3,7,8-TCDD has resulted in deaths in many species of wildlife and domestic animals. Laboratory studies with birds, mammals, aquatic organisms, and other species have demonstrated that exposure to 2,3,7,8-TCDD can result in acute and delayed mortality as well as carcinogenic, teratogenic, mutagenic, histopathologic, immunotoxic, and reproductive effects. No regulations governing PCDD contamination exist at present to protect sensitive species of wildlife and aquatic organisms. However, the limited data available suggest that 2,3,7,8-TCDD concentrations in water should not exceed 0.01 ppt to protect aquatic life, or 10 to 12 ppt in food items of birds and other wildlife (Eisler, 1986).

> *Polychlorinated dibenzo-para-dioxins (PCDDS) are present as trace impurities in some manufactured chemicals and industrial wastes.*

Sources

PCDDs are present as trace impurities in some commercial herbicides and chlorophenols. They can be formed as a result of photochemical and thermal reactions in fly ash and other incineration products.

Their presence in manufactured chemicals and industrial wastes is neither intentional nor desired. The chemical and environmental stability of PCDDs coupled with their potential to accumulate in fat has resulted in their detection throughout the global ecosystem. The number of chlorine atoms in PCDDs can vary between one and eight to produce up to 75 positional isomers. Some of these isomers are extremely toxic, while others are believed to be relatively innocuous. The most toxic and extensively studied PCDD isomer is 2,3,7,8-TCDD. PCDD-contaminated phenoxy herbicides are not the only sources of 2,3,7,8-TCDD. Others include polychlorinated biphenyls and pentachlorophenols.

Environmental Chemistry

The PCDDs consist of 75 isomers that differ in the number and position of attached chlorine atoms; each isomer has its own unique identity and toxicological properties. The most toxic of the chlorinated dioxin isomers is 2,3,7,8-TCDD. It is one of 22 possible congeners of tetrachlorodibenzo-*p*-dioxin. There is general agreement that PCDDs, including 2,3,7,8-TCDD, are (or were, until recently) found in chlorophenols, especially trichlorophenol and pentachlorophenol, in certain phenoxy pesticides (2,4,5-T; 2,4-D; Fenoprop; Silvex; Ronnel; Erbon; Agent Orange), in hexachlorophene, and in polychlorinated biphenyls (used in electrical transformers and capacitors, and contaminated with trichlorobenzenes).

PCDDs enter the environment through accidental release during chlorophenol production, through aerial application of some phenoxy herbicides, and through improper disposal of wastes into terrestrial and aquatic ecosystems (Kamrin and Rodgers, 1985). PCDDs have been identified in effluents from combustion products of municipal and industrial incinerators, including fly ash and flue gas (Czuczwa and Hites, 1984). These PCDDs may be associated with small particles, which have long residence times in the atmosphere and can become distributed over large areas. High-temperature combustion of bituminous coal in an oxidized and chlorinated atmosphere (produced experimentally) yielded chlorodioxins, mostly octa-, hepta-, and hexa-CDDs and measurable quantities of tetra-CDDs (Mahle and Whiting, 1980). Other potential sources of PCDDs include fossil fuel power plants, internal combustion engines, home fireplaces, and cigarette smoke (Kociba and Schwetz 1982).

> *The PCDDs consist of 75 isomers that differ in the number and position of attached chlorine atoms; each isomer has its own unique identity and toxicological properties.*

Bioavailability

Data on the bioavailability of PCDDs are scarce. It is known that PCDDs incorporated into wood as a result of chlorophenol (preservative) treatment are bioavailable. Toxicities of individual PCDD isomers can vary by a factor of 1,000 to 10,000 for isomers as closely related as 2,3,7,8-TCDD and 1,2,3,8-TCDD or 1,2,3,7,8-penta CDD and 1,2,4,7,8-penta CDD (Rappe, 1984). Isomers with the highest biological activity and acute toxicity have 4 to 6 chlorine atoms, and all lateral (i.e., 2,3,7, and 8) positions substituted with chlorine. On this basis, the most toxic PCDD isomers are 2,3,7,8-TCDD, 1,2,3,7,8-penta CDD, 1,2,3,6,7,8-hexa CDD, 1,2,3,7,8,9-hexa CDD, and 1,2,3,4,7,8-hexa CDD (Rappe, 1984).

Removal of Dioxins

Although PCDDs are highly persistent, volatilization and photolysis are major removal processes (NRCC 1981). In soils, 2,3,7,8-TCDD undergoes photolysis rapidly on the surface in a few hours, but more deeply buried 2,3,7,8-TCDD could have a chemical degradation greater than 10 years (NRCC 1981). Microbial degradation of 2,3,7,8-TCDD in soils is slow, with biological degradation estimated at 1.0 to 1.5 years (Ramel, 1978). In general, wherever degradation of dioxins have been detected in the environment, a local application of TCDD-contaminated herbicide, hazardous waste site, or industrial discharge has usually been implicated as the source (Stolzenburg and Sullivan, 1983). The source of these compounds is the atmospheric transport of dioxins formed by combustion of domestic and chemical wastes.

Toxic and Sublethal Effects

Information is lacking or scarce on the biological properties of PCDD isomers, except 2,3,7,8-TCDD. The latter has been associated with lethal, carcinogenic, teratogenic, reproductive, mutagenic, histopathologic, and immunotoxic effects. Animals poisoned by 2,3,7,8-TCDD-exhibit weight loss, atrophy of the thymus gland, and eventually death. Mutagenic responses were produced in *Escherichia coli* and certain strains of *Salmonella typhimurium* bacteria by 2,3,7,8-TCM,- but not by octa-CDD (Vos, 1978). Further, chromosomal aberrations were induced in at least one species of higher plants and mammals (Ramel, 1978). Sensitive species of teleosts exhibited reduced growth and fin necrosis at concentrations as low as 0.1 ppt of 2,3,7,8-TCDD after exposure for 24 to 96 hours. Histopathologic and teratogenic effects were noted in fry of rainbow trout (*Salmo gairdneri*) exposed to 10 ppt of 2,3,7,8-TCDD for 96 hours as eggs, or is yolk-sac fry (Helder, 1981). Although the mechanisms of 2,3,7,8-TCDD toxicity are unclear, current research areas include the role of thyroid hormones, interference with plasma membrane functions, alterations in ligand receptors, the causes of hypophagia (reduced desire for food) and subsequent attempts to alter or reverse the pattern of weight loss, and excretion kinetics of biotransformed metabolites.

RADIOACTIVE WASTES

Radioactive wastes can be categorized as low, intermediate and high levels. Hospitals, laboratories, and other buildings where radioactive technology is used emit low-level radiation. Intermediate level waste consists of substances from nuclear power stations like cleaning agents and sludge. This material is bulky and decays slowly. High-level nuclear wastes are extremely dangerous, that are mostly spent nuclear fuel that comes from nuclear reprocessing plants. Radionuclides enter into the environment mainly in the form of effluents produced due to nuclear power reactors (NPRs). Globally, about 440 nuclear power reactors generate 360,000 MWe. Boiling water reactors (BWRs), pressurized water reactors (PWRs) and pressurized heavy water reactors (PHWRs) generate radionuclides at all steps in the nuclear fuel cycle. Chemical approaches for the remediation of nuclear wastes are often expensive and lack specificity.

> *Radionuclides enter into the environment mainly in the form of effluents produced due to nuclear power reactors (NPRs).*

Ionizing Radiation and its Effects on Bacteria

Ionizing radiation is radiation with sufficient energy to ionize molecules. There are two types of ionizing radiation, both produced by the decay of radioactive elements: electromagnetic (X-rays and γ-radiation that form the part of electromagnetic spectrum that includes visible light and radio waves) and particulate (α and β particles). Ions react with other molecules to produce free radicals on a nanosecond timescale. In water, the products include highly reactive hydroxyl radicals. The ions and free radicals produced as radiation passes through matter react rapidly and modify molecules. Of all the effects, genome damage probably has the greatest impact on cell viability. Ionizing radiation generates multiple types of DNA damage. The damage resulting from the action of hydroxyl radicals is similar to that produced by oxidative damage associated with endogenous aerobic metabolism.

Constraints for the Bioremediation of Radionuclides

Bioremediation studies concerned with radionuclides have been confined to the laboratory due to the following reasons

1. The radiotoxicity of effluents containing high-activity radionuclides, e.g. ^{241}Pu adversely affect the microbial component of a bioprocess.

2. Most wastes from mining and fuel reprocessing contain a large background of nonfissile ^{238}U, non-penetrating α-emissions of the transuranic elements, moderately penetrating β-emission of ^{99}T$_c$ and ^{241}Pu and highly penetrating γ-rays from isotopes such as ^{60}C and ^{241}Am.

3. Wastes containing anionic species, chelating agents and citrate complex with radionuclides.

Radioresistant Bacteria

Deinococcus radiodurans is the bacterium that can resist radiation (White *et al.*, 1999). The genus name *Deinococcus* was based on the Greek adjective 'deinos', which means strange or unusual, an apt description for an organism with an ability to survive genetic damage that sets it apart from much of the life on earth. Recent studies have shown that *D. radiodurans* is able to reduce U (VI) to Te (VII) indirectly by using the electron shuttle anthraquinone-2, 6-disulphonate. This organism may be useful for *ex situ* treatment since it might not compete successfully with organisms indigenous to the subsurface. It is a non-pathogenic, soil bacterium that can grow continuously in the presence of 60 Gy/h (a dose that exceeds those in most radioactive waste sites) with no effect on either its growth rate or its ability to express foreign genes (Lange *et al.,* 1998). This organism exhibits 4–10 identical copies of a chromosome (2.65 Mbp), two megaplasmids (412 and 177 bp) and a plasmid (46 kbp). *D. radiodurans* has been engineered for resistance to mercury and organic solvents. *D. geothermalis,* an extremely- radiation resistant thermophilic bacterium, was genetically engineered for the bioremediation of radioactive waste exhibiting high-temperatures (Hassan *et al.,* 2003). Diels *et al.* (1955) reported that microbes are able to precipitate metals and radionuclides as carbonates and hydroxides via plasmid-borne resistance mechanisms, whereby influx countercurrent (antiport) to metal efflux results in localized alkalization at cell surface. Cox and Battista (2005) listed members of six bacterial phyla that are resistant to ionizing radiation (Table 7.7).

Table 7.7 Species of ionizing-radiation-resistant bacteria

Species	Phylum
Methylobacterium radiotolerans	α-Protobacteria
Kocuria rosea	Actinobacteria
Acinetobacter radioresistens	γ-Protobacteria
Kineococcus radiotolerans	Actinobacteria
Hymenobacter actinosclerus	Flexibacter, Cytophaga, Bacteroides
Chroococcidiopsis spp.	Cyanobacteria
Rubrobacter xylanophilus	Actinobacteria
Deinococcus radiodurans R1	Deinococcus, Thermus
Rubrobacter radiotolerans	Actinobacteria

> *D. geothermalis,* an extremely- radiation resistant thermophilic bacterium, was genetically engineered for the bioremediation of radioactive waste exhibiting high-temperatures.

Microbial Adaptation for Radioactivity

Microbes that are useful for bioremediation of radionuclides need to survive radiotoxicity of effluents containing high-activity radionuclides such as ^{235}U, $^{99}T_c$ and ^{241}Pu. Treatment of *D. radiodurans* with high levels of ionizing radiation can produce hundreds of genomic double-stranded breaks, but the genome is reassembled accurately before initiation of the next cycle of cell division. Studies have also shown the existence of a mechanism that blocks the formation of strand breaks in the *D. radiodurans* genome. Experimental evidences indicate that the recovery of *D. radiodurans* from substantial DNA damage relies on deinococcal physiology and a robust complement of repair enzymes. Several other mechanisms of radioresistence are as follows.

1. *Genome copy number*   Cells with increased number of genome copies have enhanced resistance to ionizing radiation. Presence of multiple genomes provide additional copies of crucial loci that improve the chance of the cell surviving irradiation. Redundant genetic information also functions as a reserve that can be used to repair DNA segments that are damaged or degraded beyond repair.

2. *Nucleoid organization*   The nucleoids of stationary phase *D. radiodurans* cells are arranged as a tightly structured ring that remains unaltered by high-dose irradiation. This structure prevents the fragmentation that is formed by double-strand breaks from diffusing apart during repair, which maintains the linear continuity of the genome even when it is fragmented (Levin-Zaidman *et al.*, 2003)

3. *Manganese content*   High concentration of Mn (II) that can accumulate in *D. radiodurans* gives the organism the capacity to survive irradiation (Daly *et al.*, 2004). Intracellular Mn (II) is protective in most DNA damage that occurs after exposure to ionizing radiation by scavenging reactive oxygen species (ROS). Daly *et al.* (2004) have proposed that Mn (II) accumulation

prevents superoxide-related ROS that are produced during irradiation from damaging proteins.

4. *Potential enzymatic contributions to repair*   A search for novel genes that are induced is response to ionizing radiation and desiccation, using genome-based microarrays, provided new evidence for both RecA-independent and RecA-dependent pathways of double-strand-break-repair (Tanaka *et al.*, 2004).

5. *DNA end protection*   Genetic analysis of radiation-resistant *D. radiodurans* revealed that two of the loci (ddrA and ddrB) contribute to radio resistance through different RecA-independent processes. The ddrA protein binds to the 3´ ends of single-stranded DNA *in vitro* and protects them from nucleolytic degradation. ddrA seems to function as a DNA-end-protection system. By protecting the broken DNA ends, cells could preserve genomic DNA until conditions become suitable for cell growth and DNA repair.

6. *RecA-independent double-strand break repair*   Levin-Zaidman *et al.* (2003) suggested that non-homologous end joining (NHEJ) would be a useful process for the repair of double-strand breaks in the context of a condensed chromosome, in which ends might not be free to diffuse away from each other.

7. *Recombinational DNA repair*   The RecA protein (361 amino acids, Mr38,013) of *D. radiodurans* promotes all of the key recombinogenic activities of RecA-class recombinases. It forms filaments on DNA, hydrolyses ATP and dATP in a DNA-dependent fashion and promotes DNA exchange (Kim *et al.*, 2002).

Bioremediation of Uranium

Role of microbes in the biogeochemical cycling of radionuclides and advances in understanding the mechanism with the activity to fine-tune their activities using the molecular tools has led to the development of novel or improved bioremediation processes (Lloyd and Lovely, 2001). Biological reduction of U (VI) to insoluble U (IV) can be stimulated by the addition of ethanol and trimetaphosphate to contaminated ground water. Sulphate-reducing bacteria (SRB) were thought to be responsible for U(VI) reduction, but a detailed analysis of the microbes involved was not known. SRB-containing microbial mat immobilized into silica pesticides can be used to clean U(VI) contaminated ground water. U(VI) reduction can be enhanced by the addition of sulphate. Studies show the involvement of the nitrite-reducing pathway in U(VI) reduction. α-emitter $^{237}N_p$ present in low-active nuclear wastes can be removed using a combination of the biological reduction of Np(V) by *S. putrefaciens* followed by precipitation of Np(IV) phosphate by a *Citrobacter* sp. Biosorption is an ideal alternative strategy for uranium removal. Biosorbents such as filamentous fungi, yeast, bacteria, actinomycetes and fresh water algae, such as chlorella have been reported binding uranium. Marine algae are capable of biologically concentrating radionuclides such as radium, thorium and uranium.

GENETICS OF MICROBIAL BIOREMEDIATION

Microbial Genetic Plasticity
Role of Plasmids in Bioremediation
Evolution Barriers for New Microbes
Enhancement of Novel Microbial Degradative Abilities
Genetics and Gene Manipulation

ENVIRONMENTAL INFORMATICS

Environmental Informatics (EI) is a new and emerging field that applies information science, engineering and technology to explain and find solutions to environmental problems. Drawing inspiration from the useful tools of the Information age, EI aims to bring about harmony between humans and nature through augmentation and amplification of the human mind, rather than its physical power. It attempts to create an easily accessible "Single umbrella" framework to establish an Environmental Information System (EIS) to pool in the large quantities of in-depth knowledge produced by environmental and related sciences. It also evaluates the EIS design for the transformation of raw environmental data into a higher grade of knowledge suitable for decision making and implements EIS using modern computers and communication systems.

Joseph and Nagendran (2004)

GENETICALLY ENGINEERED MICROBES (GEMS) IN BIOREMEDIATION

In recent years attempts have been made to create genetically engineered microbes (GEMs) to enhance bioremediation besides degrading xenobiotics which are highly resistant to breakdown (recalcitrant). The majority of the genes responsible for the synthesis of biodegradative enzymes are located on the plasmids. New bacterial strains can be created by transfer of plasmids (by conjugation) carrying genes for different degradative pathways. The following is a list of GEMs that can degrade xenobiotics.

GEMs	Xenobiotic
Pseudomonas diminuta	Parathion
P. oleovorans	Alkane
P. cepacia	2, 4, 5- Trichlorophenol
P. putida	Mono- and dichloro aromatic compounds
Alcaligenes sp.	2, 4-Dichlorophenoxy acetic acid
Acinetobacter sp.	4-chlorobenzene

Bioremediation of contaminants is an extension of normal microbial metabolism that depends on the existing genome of the microbe. Microbes face several challenges to modify the existing genome to counteract the xenobiotic compounds and direct them into central metabolic pathways. The existing genetic potential may be amplified so that metabolic processes can be accelerated to degrade new compounds. The study of the genetic mechanisms for change is essential to evaluate the degradative capacity of microbes.

MICROBIAL GENETIC PLASTICITY

Evolution of new metabolic potential to degrade xenobiotics depends on the susceptibility to alteration and exchange of genetic information. Mutations range from single-nucleotide insertions, deletions or substitutions to larger-scale sequence requirements. Errors in DNA replication or repair can result in single-nucleotide mutations. Gene conversion, gene duplication and transposition play important roles in the rearrangement of DNA fragments and in the activation or inactivation of cryptic genes (Vandermeer *et al.*, 1992). Microbes may also acquire or lose genetic material by interactions with other organisms during conjugation, transformation and transduction. Plasmids and transposons play important roles in shuttling genetic information from host cells to recipients (Ravora, 1989). Documentation of gene flow in the field is difficult because of the paucity of readily identifiable genetic markers and the inherent intractability of the natural ecosystem to controlled study. Microbial communities are able to adapt to a variety of contaminants due to mechanisms similar to selective antibody-resistance. Adaptive responses involve the natural selection of mutants possessing degradative enzymes with relaxed substrate specificities or even new metabolic activities. Protein sequence analysis will provide evidence for genetic mechanism of adaptation (Harayama *et al.*, 1987).

> *The existing genetic potential may be amplified so that metabolic processes can be accelerated to degrade new compounds.*

> *Plasmids and transposons play important roles in shuttling genetic information from host cells to recipients.*

ROLE OF PLASMIDS IN BIOREMEDIATION

The nature of the environment dictates the mode of evolution and the development of new degradative enzyme genes due to the constant pressure of specific substrate conditions. Presence of catabolic plasmids among microbial communities is believed to be an important mechanism for genetic adaptation. Many plasmid-containing bacteria have been isolated from polluted areas, which are relative to those present in pristine control sites. Plasmids are important in the rapid transfer of genes among populations in a microbial community. Hospital wastes, raw sewage, sewage effluents, fresh and marine water, animal feedlots, plants and soils have all been shown to contain bacteria that transfer plasmids by conjugation. Plasmids with antibiotic resistance genes, as well as those with degradative pathways, can be transferred to a wide variety of bacterial species in many genera. Bacteria like *Pseudomonas* have plasmids which code for degradative enzymes. Complete degradation of growth substrate is possible with the help of such plasmids alone (Table 8.1). The transfer of plasmids to different strains ensures transfer of complete degradative ability to recipient for the concerned growth substrate. Recipient with its own chromosomal enzymes may metabolize it further.

Table 8.1   Types of degradative plasmids

Plasmid	Substrate	Plasmid coded path way
CAM	Camphor	Camphor→Acetate + Isobutyrate
OCT	Octane, hexane, decane	Octane→Octanoate
TOL	Toluene, nuta-xylene paraxylene	Toluene→Benzoate→Catechol
NAH	Naphthalene	Naphthalene→Salicylate→Catechol– Acetaldehyde + Pyruvate
SAL	Salicylate	Salicylate→Catechol

> *The transfer of plasmids to different strains ensures transfer of complete degradative ability to recipient for the concerned growth substrate.*

Biodegradation can be augmented by the addition of surfactants, supplementation with inorganic nutrients and inoculation with biomass of enriched bacterial species. The enzymes responsible for biodegradation are genetically encoded in bacterial plasmids. These plasmids possess broad host range and can be transmitted within the same species or genera. They have been identified in species of *Pseudomonas, Alcaligens, Acinetobacter, Flavobacterium, Beijerinkia, Klebsiella, Moraxella,* and *Arthrobacter.* Most of the novel strains have been genetically engineered from the genes of *Pseudomonas* and have been patented.

Microbial communities exposed to hydrocarbons become adapted, exhibiting selective enrichment and genetic changes resulting in increased proportions of hydrocarbon-degrading bacteria and bacterial plasmids encoding hydrocarbon catabolic enzymes (Colwell, 1990). Studies in *Plasmodium aeruginosa* containing OCT plasmid showed that a number of loci on the plasmid and chromosome are involved in the coding of the enzymes of the degradative pathway. In bacteria, the OCT plasmid is the only well-characterized system, which codes for a number of proteins involved in the growth of bacteria on *n*-alkanes ranging from C_6 to C_{10}, In *P. putida* P_6G_6 the metabolic pathway involved in *n*-alkane degradation is alkane hydroxylase (monooxygenase). The resultant product (alcohol) is further oxidized by the alcohol dehydrogenase as well as aldehyde dehydrogenase to yield fatty acids. These fatty acids are further metabolized through the β-oxidation pathway. Exchange of genetic information between the indigenous microbiota and plasmid-carrying strains of *P. aeruginosa* and *P. putida* has resulted in the appearance of a bacterial population capable of using 3-chlorobenzoate as a sole source of carbon and energy (Pertson, *et al.,* 1984).

> *Biodegradation can be augmented by the addition of surfactants, supplementation with inorganic nutrients and inoculation with biomass of enriched bacterial species.*

> *In bacteria, the OCT plasmid is the only well-characterized system, which codes for a number of proteins involved in the growth of bacteria on n-alkanes ranging from C_6 to C_{10}.*

EVOLUTION BARRIERS FOR NEW MICROBES

New microbes are unable to evolve, for the degradation of a myriad of chemicals in the environment due to several reasons. The multiple reasons are:

1. The contaminants are often found in low concentrations that are insufficient to sustain the growth of microbes that are capable of metabolizing them.

2. There is only low selective pressure to enrich microbes that have the capacity to metabolize the pollutants.

3. There exists a fluctuation in the levels of contaminants.

4. The toxicity of contaminants may kill the entire microbial population before the appearance of mutants.

5. Many of the chemical contaminants have been introduced relatively recently and the microorganisms have not yet had enough time to evolve metabolic pathways for their degradation.

ENHANCEMENT OF NOVEL MICROBIAL DEGRADATIVE ABILITIES

Random mutation and environmental selection represent the primary mechanisms for the evolution of the ability to metabolize xenobiotic substrates in bacteria. Adaptive features introduced into the microbial communities allow for the quick and widespread expression of the features due to their rapid reproductive rates. Persistence of any characteristic in a population depends on its genetic fitness, that is the contribution of one or more gene alleles of the population to succeeding generations (Lenski, 1992). Degradative ability of xenobiotics by microbes depends on genetic rearrangement or exchange. It is related to the frequency of gene replication and inter-organism interactions. High growth rates and dense microbial populations favouring such interaction are not usually encountered under field conditions. Exploitation of the natural evolutionary process to develop organisms with novel degradative abilities can be an extremely slow and unpredictable process. Hence construction of suitable organisms by genetic manipulation is a feasible alternative, with the following characteristics:

1. expansion of substrate range of existing pathways through the recruitment of isofunctional enzymes (horizontal expansion) and grafting of additional activities (vertical expansion)

2. existing pathways can be restructured to avoid the generation of deleterious metabolites

3. construction of new metabolic activities by the patchwork assembly of genes from different microbes that encode desirable enzymatic conversions

4. new sequence construction by enhancing natural genetic exchanges among dissimilar microbes with imposed selection pressure (molecular breeding)

> *Degradative ability of xenobiotics by microbes depends on genetic rearrangement or exchange.*

GENETICS AND GENE MANIPULATION

VazQuez–Duhalt (2003) categorized different genetic mechanisms of adaptation to metabolize xenobiotic substrates: gene transfer, and genetic mutation that includes DNA rearrangement, gene duplication, transposition and activation by insertion. Three principal mechanisms of genetic transfer and recombination lead to new combinations of alleles among bacteria. They are

1. *Transformation*   a process involving DNA uptake;

2. *Conjugation*   a process that involves direct contact between donor and recipient cells and

3. *Transduction*   a process that involves bacteriophage-mediated transfer of DNA from donor to recipient bacterial cell.

There is a relatively high potential for gene transfer by these mechanisms in the environment, particularly when population densities are high.

Endeavours are being made to practically modify microbes into more efficient degraders using modern tools of molecular biology and metabolic engineering. Genes can be transferred to new populations within the community to form new allelic combinations with differing degrees of fitness. *In vivo* gene manipulation involves the usual genetic transfer mechanisms, while *in vitro* techniques expand the natural limits of gene alteration. *In vivo* mechanisms are time-consuming and unpredictable, but the resulting microbes are genetically altered to a more limited extent. *In vitro* gene manipulation can be used to construct organisms with more extensive alterations and this is relatively fast, efficient and predictable. However this technique is subject to a variety of societal concerns.

Metagenomics in Bioremediation

To analyse the diversity of pollutant-degrading population in the environment, fragments of genes for catabolic enzymes in degradation pathways/16S rDNA have been amplified from environmental DNA samples using polymerase chain reaction (Lorenz *et al.,* 2004). Since, this approach requires prior information of the gene sequences for designing suitable primers, metagenomics offers the advantage of direct cloning of soil metagenomic DNA. Metagenomics constitutes the collective genome of soil microflora. Daniel (2004) reported that this approach can be successfully utilized for accessing the biosynthetic diversity of microbes from the environment by acquiring operons or gene(s) involved in synthesis of beneficial compounds or in biodegradation of recalcitrant pollutants. As the genes encoding the enzymes of a degradation pathway are usually clustered in a contiguous piece of DNA, large fragments of DNA are usually cloned in Bacterial Artificial Chromosome (BAC), fosmid or cosmid vectors such that the entire metagenomic library is represented by a finite number of screenable clones. Henne *et al.* (1999) characterized the enzyme *p*-hydroxybutyrate dehydrogenase involved in degradation of *p*-hydroxybutyrate using metagenomic profiling. This indicates that metagenomic profiling promises an enormous possibility of identifying novel enzymes or pathways involved in biodegradation of poorly degraded pollutants. Metagenomics can be broadened by introducing the desired gene(s) into other hosts such as *Pseudomonas.*

Future Perspectives

The advanced culture-independent techniques of microbial community analysis, including all their limitations and bases, seem to be suitable for determination of virtual changes in the community structure exposed to the impact of pollutants (Kozdroj and Elsas, 2001). Development of genetically engineered microorganisms (GEMs) or "designer biocatalysts" harbouring artificially designed metabolic pathway can overcome the limitation of bioremediation. Databases such as the "University of Minnesota biocatalysis/ Biodegradation database (http://umbbd.ahc.umn.edu) and "Biodegradative strain database (http://bsd.cme.msu.edu/bsd/ index.html) provide a scope for *in silico* designing of "biocatalysts" for *in vivo* construction followed by *in situ* application. The databases generated by The Institute of Genomic Research (TIGR) and Monterey Bay Coastal Ocean Microbial Observatory (http:// www.tigr.org./tdb/mbmo/) would expand the available metagenomic information about some of the more obscure microbial metabolic potentials. Perhaps the evolving new branch of science, nanobiotechnology involving the use of bio-microelectronics, micromachines and exploitation of the information processing abilities of microbial cells, may solve the constraints of microbial bioremediation.

GLOSSARY

A horizon   The uppermost layer of soil; topsoil.

Abiotic   The absence of living organisms (non-living).

Absorption   The uptake, drinking in, or imbibing of a substance; the movement of substances into cells; the transfer of substances from one medium to another, e.g. the dissolution of gas in a liquid; the transfer of energy from electromagnetic waves to chemical bond and/or kinetic energy, e.g. the transfer of light energy to chlorophyll.

Accelerated   bioremediation   Bioremediation within the subsurface at a given site that is accelerated beyond the normal actions of the naturally occurring microbial community and naturally occurring chemical and geological conditions.

Acid mine drainage   Consequences of the metabolism of suphur and iron-reducing bacteria when coal mining exposes pyrite to atmospheric oxygen and the combination of auto-oxidation and microbial sulphur and iron oxidation produces large amounts of sulphuric acid, which kills aquatic life and contaminates water.

Acidic   A compound that releases hydrogen (H^+) ions when dissolved in water; a compound that yields positive ions upon dissolution; a solution with a pH value less than 7.0.

Acidophile   Organism that grows best under acid conditions (down to a pH of 1).

Acridine Orange Direct Count (AODC)   A direct count method using the fluorescent dye acridine orange to stain bacterial cells; it detects numbers of cells, both living and dead.

Activated sludge process   An aerobic secondary sewage treatment process using sewage sludge containing active complex populations of aerobic microbes to break down organic matter in sewage.

Adaptation   Change in an organism or population of organisms through which they become more suited to the prevailing environment. Adaptation can be genetic and/or physiological.

Adsorption coefficient (K_{oc})   The ratio of the amount of a chemical adsorbed per unit weight of organic carbon in the soil or sediment to the concentration of the chemical in solution at equilibrium.

Adsorption ratio (Kd)   The amount of a chemical adsorbed by a sediment or soil. (i.e., the solid phase) divided by the amount of chemical in the solution phase, which is in equilibrium with the solid phase, at a fixed solid/solution ratio. It is generally expressed in micrograms of chemical sorbed per gram of soil or sediment.

Adsorption   A surface phenomenon involving the retention of solid, liquid, or gaseous molecules at an interface.

Advection   The process by which solutes are transported by the bulk motion of the flowing ground water.

Aerated pile method Method of composting for the decomposition of organic waste material where the wastes are heaped in piles and forced aeration provides oxygen.

Aerobe An organism that can grow in the presence of air.

Aerobic An environment that has a partial pressure of oxygen (molecular) similar to normal atmospheric conditions.

Alkalophile Organism that grows best under alkaline conditions (up to a pH of 10.5).

Allochthonous An organism or substance foreign to a given ecosystem.

Amensalism An interspecific interaction in which one species population is inhibited, typically by the toxin produced by the other, which is unaffected.

Ammonification The release of ammonia from nitrogenous organic matter by microbial action.

Anaerobe An organism that grows in the absence of oxygen or air.

Anaerobic digester A secondary sewage treatment facility used for the degradation of sludge and sewage waste.

Anaerobic respiration Use of inorganic electron acceptors other than oxygen as terminal electron acceptors for energy-yielding oxidative metabolism. Nitrate respiration is an example of anaerobic respiration.

Anaerobic An environment without oxygen.

Anoxic Literally "without oxygen." An adjective describing a habitat devoid of oxygen.

Anoxyphotobacteria (Anaerobic photosynthetic bacteria) Bacteria that have only photosystem I and do not evolve oxygen in the course of their photosynthesis. They live in anaerobic aquatic habitats that receive some light.

Anthropogenic Derived from human activities.

Aquifer A geological formation containing water, such as surface water bodies that supply the water for wells and springs; a permeable layer of rock or soil that holds and transmits water.

Autochthonous Microbes/ substances indigenous to a given ecosystem; the inhabitants of an ecosystem, referring to the common microbiotas of the body or soil microorganisms that tend to remain constant despite fluctuations in the quantity of fermentable organic matter.

Autotrophs Organisms whose growth and reproduction are independent of external sources of organic compounds. The required cellular carbon is supplied by the reduction of CO_2 and the needed cellular energy being supplied by the conversion of light energy to ATP or the oxidation of inorganic compounds to provide the free energy for the formation of ATP.

B horizon The soil layer beneath the A horizon, consisting of weathered material and minerals leached from the overlying soil.

Bacteria A group of diverse and ubiquitous prokaryotic single-celled microorganisms.

Bioaccumulation Intracellular accumulation of environmental pollutants such as heavy metals by living organisms.

Bioaugmentation The addition to the environment of microorganisms that can metabolize and grow on specific organic compounds.

Bioavailability The availability of chemicals to microbes that can degrade them.

Bio-barrier A biologically active zone that is placed in the subsurface perpendicular to the normal flow of a contaminant plume so that the contaminant can be adsorbed and biologically degraded.

Bio-chelator A biochemical compound (synthesized by living organisms) that binds and forms complexes with trace elements and polyvalent cations.

Biochemical oxygen demand (BOD) The requirement for molecular oxygen by microbes during oxidation of biological substances in sewage. The BOD test measures the oxygen consumed (in mg/L) over 5 days at 20°C.

Bioconcentration factor (BC) The quotient of the concentration of a chemical in aquatic organisms at a specific time or during a discrete time period of exposure divided by the concentration in the surrounding water at the same time during the same period.

Biodegradation The breakdown of organic substances by microorganisms or their enzymes.

Biodeterioration The chemical or physical alteration of a product that decreases the usefulness of that product for its intended purpose.

Biodisc system A secondary sewage treatment system employing a film of active microbes rotated on a disc through sewage.

Biofilm A microbial community occurring on a surface as microlayer.

Biofilter Apparatus that biodegrades volatile organic compounds in air by passing the air through media containing biodegrading microbes.

Biogenic element An element that is incorporated into the biomass of living organisms.

Biogeochemical dynamics The temporal behaviour of coupled biological, hydrological, and geochemical processes in geological media.

Bioleaching The use of microbes to transform elements so that the elements can be extracted from a material when water is filtered through it.

Biomass The amount of living matter present in a particular habitat.

Biopolymer Microbial products, either exopolysaccharides or polyhydroxyalkanoates, that can be manufactured commercially from renewable resources, and that are biodegradable and provide alternatives to traditional plant and algal gums or to plastics made from hydrocarbons.

Bioremediation   The process by which living organisms act to degrade or transform hazardous organic contaminants.

Bioscrubber   A device in which air moves through a fine spray or a microbial suspension in order to remove pollutants from air.

Biosequestration   The conversion of a compound through biological processes to a form that is chemically or physically isolated.

Biostimulation   A process that increases activity of microorganisms biodegrading contaminants. e.g. addition of nutrients, oxygen, or other electron donors and acceptors.

Biosurfactant   A surface-active agent produced by microbes.

Biotic   Living.

Biotransformation   Alteration of the structure of a compound by a living organism or enzyme.

Bioventing   The process of supplying oxygen *in situ* to oxygen-deprived soil microbes by forcing air through unsaturated contaminated soil at low flow rates. This stimulates biodegradation and minimizes stripping volatiles into the atmosphere. Frequently used to remediate soil under structures since it is relatively non-invasive.

Black smoker   A geothermal eruption occurring in deep-sea regions that emits metal- and sulphide-rich waters that form black metal precipitates.

Bloom   A variable abundance of microbes, generally referring to the excessive growth of algae or cyanobacteria at the surface of a body of water.

Bottle effect   Growth of microbes within a collection vessel that results in artificially elevated microbial counts.

Breakpoint chlorination   Procedure for the removal and oxidation of ammonia from sewage to molecular nitrogen by the addition of hypochlorous acid.

Brownfield   An abandoned, idled, or under-used industrial or commercial facility where expansion or redevelopment is complicated by a real or perceived environmental contamination.

BTEX   Benzene, Toluene, Ethylbenzene and Xylene.

C horizon   The soil layer beneath the B horizon consisting of the broken or partially decomposed underlying bedrock.

Catabolic pathway   A degradative metabolic pathway; a metabolic pathway in which large molecules are broken down into smaller ones.

Ceiling value   The concentration of a substance that should not be exceeded, even instantaneously.

Chemical Oxygen Demand (COD)   The amount of oxygen required in milligrams per litre to oxidize both organic and oxidizable inorganic compounds.

Chemoautotrophs Microorganisms that obtain energy from the oxidation of inorganic compounds and carbon from inorganic carbon dioxide; organisms that obtain energy through chemical oxidation and use inorganic compounds as electron donors; also known as chemolithotrophs.

Chemolithotrophs Microorganisms that obtain energy through chemical oxidation and use inorganic compounds as electron donors and cellular carbon through the reduction of carbon dioxide; also known as chemoautotrophs.

Chemoorganotrophs Organisms that obtain energy from the oxidation of organic compounds and cellular carbon from preformed organic compounds.

Colony hybridization Hybridization that is combined with conventional plating procedures in which bacterial colonies or phage plaques are transferred directly onto hybridization filters; the colonies or phage-containing plaques are then lysed by alkaline or enzymatic treatment, after which hybridization is conducted.

Colony-forming units (CFUs) Number of microbes that can replicate to form colonies, as determined by the number of colonies that develop.

Co-metabolism The biodegradation of a pollutant by an organism while using some other compound(s) for growth and energy. There is little or no benefit to the biodegrading organism, the pollutant just happens to be affected by the growth of the co-metabolizing organism.

Commensalism An interactive association between two populations of different species living together, in which one population benefits from the association, and the other is not affected.

Competition An interactive association between two species, both of which need some limited environmental factor for growth and thus grow at sub-optimal rates because they must share the growth-limiting resource.

Competitive exclusion principle The principle that states that competitive interactions tend to bring about the ecological separation of closely related populations and preclude two populations from occupying the same ecological niche.

Competitive inhibition The inhibition of enzyme activity caused by the competition of an inhibitor with a substrate for the active (catalytic) site on the enzyme; impairment of the function of an enzyme due to its reaction with a substance chemically related to its normal substrate.

Composting The decomposition of organic matter in a heap by micro-organisms; a method of solid waste disposal.

Consortium Two or more members of a natural assemblage of microbes in which each organism benefits from the other. The group may collectively carry out some process that no single member can accomplish on its own.

Coprophagous Capable of growth on faecal matter; feeding on dung or excrement.

Creosote An antifungal wood preservative used frequently to treat telephone poles and railroad ties. Creosote consists of coal tar distillation products, including phenols and PAHs.

Decomposers Organisms, often bacteria or fungi, in a community that convert dead organic matter into inorganic nutrients.

Dendrograms Graphic representations of taxonomic analyses, showing the relationships between the organisms examined.

Denitrification The formation of gaseous nitrogen and/or oxides of nitrogen from nitrate or nitrite by some bacteria during anaerobic respiration. Denitrification occurs only in anaerobic or microaerophilic conditions. It can sometimes be used to remove nitrate or nitrite from liquid wastes.

Desulphurization Removal of sulphur from organic compounds.

Diatomaceous earth A silicaceous material composed largely of fossil diatoms, used in microbiological filters and industrial processes.

Direct counting procedures Methods for the enumeration of bacteria and other microbes that do not require the growth of cells in culture but rather rely upon direct observation or other detection methods by which the undivided microbial cells can be counted.

Direct viability count A direct microscopic assay that determines whether microorganisms are metabolically active, i.e., viable.

Domestic sewage Household liquid wastes.

Effluent The liquid discharge from sewage treatment and industrial plants.

E_h (redox potential) Measure of the tendency of a given system to donate electrons (i.e., to act as a reducing agent) or to accept electrons (i.e., to act as an oxidizing agent); the E_h of a given system (measured as volts) may be determined by measuring the electrical potential difference between that system and a standard hydrogen electrode whose potential is, by convention, arbitrarily taken to be zero volts.

Electron acceptor Small inorganic or organic compound that is reduced to complete an electron transport chain. Compound that is reduced in a metabolic redox reaction.

Electron donor Small inorganic or organic compound that is oxidized to initiate an electron transport chain. Compound from which electrons are derived in a metabolic redox reaction.

Endogenous Growing from or on the inside; developing within the cell. Constituting or relating to metabolism of the nitrogenous constituents of cells or tissues.

Enhanced rhizosphere biodegradation Enhanced biodegradation of contaminants near plant roots where compounds exuded by the roots increase microbial biodegradation activity. Other plant processes such as water uptake by the plant roots can enhance biodegradation by drawing contaminants to the root zone.

Enrichment Culture in a liquid medium that results in the increase of the population of an organism relative to others. The liquid culture frequently contains substances that encourage the growth of the selected organism (such as the chemical pollutant and mineral nutrients).

Enrichment culture Any form of culture in a liquid medium that results in an increase in a given type of organism while minimizing the growth of any other organism present.

Eurythermal Microorganisms that grow over a wide range of temperatures.

Eutrophication The enrichment of natural waters with inorganic materials, especially nitrogen and phosphorous compounds that supports the excessive growth of photosynthetic organisms (plants and algae).

Ex situ Out of the original position (excavated).

Extreme environment Environment characterized by extremes in growth conditions, including temperature, salinity, pH, and water availability, among others.

Extreme thermophiles Organisms having an optimum growth temperature above 80°C.

Extremophiles A group of organisms whose growth is dependent on extreme environmental conditions, e.g. extreme halophiles, extreme psychrophiles, extreme thermophiles.

Facultative organism Organism that can carry out both options of a mutually exclusive process (e.g. aerobic and anaerobic metabolism).

Fastidious An organism with stringent physiological requirements for growth and survival; an organism difficult to isolate or culture on ordinary media because of its need for special nutritional factors.

Fermentation An energy-yielding metabolism that involves a series of oxidation–reduction reactions in which the substrate and terminal electron acceptor are organic compounds. Fermentation occurs in a wide variety of bacteria and fungi.

Floc A mass of microorganisms cemented together in a slime produced by certain bacteria, usually found in waste treatment plants.

Fungi A group of diverse, unicellular and multicellular eukaryotic organisms, lacking chlorophyll, often filamentous and spore-producing. Some species are important in the decomposition of plant litter.

Gnotobiotic Culture or environment containing only defined forms of life.

Halophiles Organisms requiring NaCl for growth; extreme halophiles grow in concentrated brines.

Hazardous chemicals   Under the OSH (Occupational Safety and Health) Act, all chemicals listed by OSHA (Occupational Safety and Health Administration) with a permissible exposure limit (PEL) by the American Conference of Governmental Industrial Hygienists (ACGH) with a Threshold Limit Value (TLV). The other hazardous chemicals include those listed in the National Toxicology Program (NTP), Annual Report on Carcinogens (latest editon) or have been found to be a potential carcinogen in the International Agency for Research on Cancer (IARC), Monographs (latest edition), or by OSHA, e.g. Tobacco or tobacco products; wood or wood products; food, drugs or cosmetics that are indented for personal use.

Hazardous material   Under DOT (Department of Transportation) rules, a substance or material which has been determined by the secretary of transformation to be capable of posing an unreasonable risk to health, safety and property when transported in commerce, and which has been so designated. The hazard categories include explosives, gases, flammable liquids, flammable solids, spontaneous combustibles and substances dangerous when wet, oxides and organic peroxides, poisonous and infectious substances, corrosives and other hazardous materials.

Hazardous samples   Samples that present a relatively high risk to health or the environment because of the potential level and type of contamination. They can be classified as high and moderate hazardous samples. The determination of the sample type is based on existing analytical data and background information.

Hazardous substances   Under the comprehensive, Environmental Response, Composition and Liability Act, elements, compounds, mixtures, solutions and substances which when released into the environment may present substantial danger to public health and welfare or the environment. Substances listed in EPA's (Environmental Production Agency) regulation 40 CFR (Code of Federal Regulations) 302.4 under CERCLA (Comprehensive Environmental Response, Compensation, and Liability Act), specifying reportable quantities (RQs) for each substance.

Hazardous wastes   Under RCRA and defined in 40CFR 261, a solid waste that, because of quantity, concentration, or physical, chemical, or infectious characteristics: (a) Causes or significantly increases mortality or serious irreversible or incapacitating reversible illness or (b) poses a substantial present or potential hazard to human health or the environment when improperly managed.

Heavy metals   Metallic elements with high molecular weights, generally toxic in low concentrations to plant and animal life. Such metals are often residual in the environment and exhibit biological accumulation. Examples include mercury, chromium, cadmium, arsenic, and lead.

Herbicides   Chemicals used to kill weeds.

Heterotrophs   Organisms requiring organic compounds for growth and reproduction; the organic compounds serve as sources of carbon and energy.

Holdfast A structure that allows certain algae and bacteria to remain attached to the substratum.

Humic acids High-molecular-weight irregular organic polymers which are acidic. The portion of soil organic matter soluble in alkali but not in acid.

Humus The organic portion of the soil remaining after microbial decomposition.

Hydrocarbons Strictly, any of a large class of organic compounds containing only carbon and hydrogen; often used to include substituted hydrocarbons (e.g. alcohols, chlorinated solvents).

Immediately Dangerous to Life or Health (IDLH) The maximum environmental concentration of a contaminant from which one could escape within 30 min. without any escape-imparing symptoms or irreversible health effects.

Immobilization The binding of a substance so that it is no longer reactive or able to circulate freely.

In situ In place, without excavation.

Indicator organism An organism used to identify a particular condition, such as *Escherichia coli* as an indicator of faecal contamination.

Indigenous Native to a particular habitat.

Inoculum Material used to introduce a microorganism into a suitable situation for growth.

Insecticides Substances destructive to insects; chemicals used to control insect populations.

Ionizing radiation Radiation such as gamma and X-radiation, that form toxic free radicals disruptive to the biochemical organization of cells.

Isoenzyme An enzyme that occurs in more than one form in a given species. Sometimes called an isozyme.

Ketone Organic compound characterized by a carbonyl group attached to two carbon atoms, usually contained in hydrocarbon radicals or in a single bivalent radical, similar to an aldehyde but less reactive.

Landfarming (land treatment) The application of toxic organic wastes to soils for the purpose of biodegradation.

Landfill A site where solid waste is dumped and allowed to decompose; a process in which solid waste containing both organic and inorganic material is deposited and covered with soil.

Leaching The process or an instance of separating the soluble components from some material by percolation. The process or an instance of removing nutritive or harmful elements from soil by percolation.

Lignin A complex polymer that occurs in woody material of higher plants. It is highly resistant to chemical and enzymatic degradation.

Liquid wastes Waste material in liquid form, the result of agricultural, industrial, and all other human activities.

Magnetotaxis Motility directed by a geomagnetic field.

Medium Any material that supports growth of an organism.

Mesocosm A field-scale model utilized to understand the interactive relationships between microbial populations and the roles played by microbial populations within ecosystems.

Mesophiles Organisms whose optimum growth is in the temperature range of 20–45°C.

Methanogens Bacteria that anaerobically oxidize hydrogen to methane and water using carbon dioxide as the electron acceptor. These occur in anaerobic muds, ponds, and sewage sludge.

Methanotroph Aerobic bacteria that can use methane as a sole source of carbon.

Methylation The process of substituting a methyl group for a hydrogen atom.

Microaerophilic An environment that is low in oxygen but is not anaerobic.

Microbial ecology The field of study that examines the interactions of microorganisms with their biotic and abiotic surroundings.

Microbial mining A mineral recovery method that uses bioleaching to recover metals from ores not suitable for direct smelting.

Microbial pesticides Preparations of pathogenic or predatory microorganisms that are antagonistic toward a particular pest population.

Microcosm A community or other unit that is representative of a larger unity. A laboratory-scale model used to understand the interactive relationships between microbial populations and the roles played by microbial populations within ecosystems.

Microflora All of the microorganisms associated with a location or environment.

Micronutrient Chemical element necessary for growth found in small amounts, usually < 100 mg kg-1 in a plant. These elements consist of B, Cl, Cu, Fe, Mn, Mo, and Zn.

Mineralization The breakdown of organic matter to inorganic materials (such as carbon dioxide and water) by bacteria and fungi.

Minimal medium Culture medium that lacks certain growth factors so that it will support growth of only certain types of microorganisms.

Mixotrophs Organisms capable of utilizing both autotrophic and heterotrophic metabolic processes, e.g. the concomitant use of organic compounds as sources of carbon, and of light as a source of energy.

Most Probable Number (MPN) A method for estimating the concentration of microorganisms in a sample. A given volume of liquid or suspension is inoculated into each of (typically) 5 tubes containing growth media. Decreasing volumes are inoculated into successive sets of 5 tubes. After an incubation period the tubes are scored for growth or lack of growth. Those tubes in which growth occurred are assumed to have contained at least one viable organism in the inoculant. The concentration of viable microorganisms in the original liquid or suspension is calculated using a statistical table.

Mycelium (plural: Mycelia) Mass of hyphae that form the vegetative body of many fungal organisms.

Mycobacterium A genus of aerobic bacteria found in soil and water that are capable of biodegrading multi-ring compounds such as PAHs.

Mycorrhiza A mutually beneficial association between a fungus and the root of a plant. These occur in a wide range of plants including trees, shrubs, and herbaceous plants.

NAPL Non-aqueous phase liquid. This can be either lighter than Non-aqueous phase liquid (LNAPL), or denser than Non-aqueous phase liquid (DNAPL).

Natural bioremediation Bioremediation at a given site as a function of the naturally occurring microbial population and naturally occurring chemical, biological, and geological conditions.

Neuston The layer of organisms growing at the interface between air and water.

Neutralism The relationship between two different microbial populations characterized by the lack of any recognizable interaction.

Niche The functional role of an organism within as ecosystem; the combined description of the physical habitat, functional role, and interactions of the microorganisms occurring at a given location.

Nitrate respiration (dissimilatory nitrate reduction) The use of nitrate as a terminal electron acceptor for anaerobic respiration. This process occurs under anaerobic or microaerophilic conditions. Not all bacteria are capable of this form of metabolism and the nitrate may not be reduced completely to nitrogen gas (stopping at nitrite, for example). When the nitrate is reduced to gaseous forms, the process is called denitrification. This can sometimes be used to remove nitrate or nitrite from liquid wastes.

Nitrification The process in which ammonia is oxidized to nitrite and nitrite to nitrate; a process primarily carried out by the strictly aerobic, chemolithotrophic bacteria of the family Nitrobacteraceae.

Nitrifying bacteria Nitro-bacteraceae; gram-negative, obligately aerobic, chemolithotrophic bacteria occurring in aquatic environments and in soil, which oxidize ammonia to nitrite or nitrite to nitrate.

Nitrite ammonification Reduction of nitrite to ammonium ions by bacteria; does not remove nitrogen from the soil.

Nitrogen fixation   The reduction of gaseous nitrogen to ammonia, carried out by certain prokaryotes.

Nitrogenase   The enzyme that catalyses biological nitrogen fixation.

No-Observed-Adverse-Effect Level (NOAEL)   The dose of chemical at which there were no statistically or biologically significant increases in frequency or severity of adverse effects seen between the exposed population and its appropriate control. Effects may be produced at this dose, but they are not considered to be adverse.

O horizon   The organic layer of soil, consisting of humic substances.

Obligate aerobes   Organisms that grow only under aerobic conditions, i.e., in the presence of air or oxygen.

Obligate   Any state or condition that is an essential attribute of a given organism, e.g. an obligate aerobe can grow only under aerobic conditions.

Octanol/Water partition coefficient (K_{ow})   The equilibrium ratio of the concentration of material partitioned between octanol and water. This coefficient is considered to be an index of the potential of a chemical to be bioaccumulated. Higher values of K_{ow} are associated with greater bioaccumulating potential.

Oligotrophic   Bodies of water poor in those nutrients that support growth of aerobic photosynthetic organisms.

Organic pump   Uptake of large quantities of water by plant (trees) roots and translocation into the atmosphere to reduce flow of water. Used to keep contaminated ground water from reaching a body of water, or to keep surface water from seeping into a capped landfill and forming leachate.

Oxidase   An enzyme that catalyses a reaction in which electrons are removed from a substrate and donated directly to molecular oxygen.

Oxidation pond   A method of aerobic waste disposal employing biodegradation by aerobic and facultative microorganisms growing in a standing water body.

Oxidation–reduction potential   A measure of the tendency of a given oxidation–reduction system to donate electrons, i.e., to behave as a reducing agent, or to accept electrons, i.e., to act as an oxidizing agent; determined by measuring the electrical potential difference between the given system and a standard reference system.

Oxygen tension   Concentration of oxygen in water.

Oxygenase   An enzyme that catalyses a reaction in which one (mono-oxygenase) or both (dioxygenase) atoms of molecular oxygen are incorporated into a molecule of substrate. Oxygenases catalyse the first step in degradation of straight-chained and aromatic hydrocarbons.

P/R ratio   The relationship between gross photosynthesis and rate of community respiration.

PAH   Polynuclear aromatic hydrocarbon. Multi-ring compounds found in fuels, oils, and creosote. These are also common combustion products.

Pathogen An organism capable of causing disease.

Pathway engineering The application of engineering technology to alter or construct a sequence of enzymatically catalysed chemical reactions.

Permissible Exposure Limit (PEL) An allowable exposure level in workplace air averaged over an 8-hour shift.

Pesticides Substances destructive to pests, especially insects.

pH The symbol used to express the hydrogen ion concentration, signifying the logarithm to the base 10 of the reciprocal of the hydrogen ion concentration.

Phenol Carbolic acid (C_6H_5OH). Phenols and substituted phenols are used as antimicrobial agents in high concentrations.

Photoautotrophs Organisms that use light as the source of energy and whose source of carbon is carbon dioxide; characteristic of plants, algae, and some prokaryotes.

Photoheterotrophs Organisms that obtain energy from light but require exogenous organic compounds for growth.

Photosynthesis The process in which radiant (light) energy is absorbed by specialized pigments of a cell and is subsequently converted to chemical energy; the ATP formed in the light reactions is used to drive the fixation of carbon dioxide, with the production of organic matter.

Photosystem I Cyclic photo-phosphorylation; photosynthetic system for generating a proton motive force, which does not require an external electron donor and does not generate reduced coenzyme.

Photosystem II Non-cyclic photo-phosphorylation; photosynthetic system for generating a proton motive force which requires an external electron donor and generates reduced coenzyme.

Phthalate A salt or ester of phthalic acid.

Plasmid Extra DNA in a cell that is usually dispensable, but may confer an advantage to the cell, such as the ability to biodegrade certain compounds or resistance to antibiotics.

Psychrophile An organism that has an optimum growth temperature below $20^\circ C$.

Psychrotroph A mesophile that can grow at low temperatures.

Putrefaction The microbial breakdown of protein under anaerobic conditions.

Radionuclides Nuclides that emit radioactivity (alpha, beta, or gamma particles) or fission into smaller nuclides.

Recalcitrant A chemical that is totally resistant to microbial attack.

Redox potential The oxidation–reduction potential of an environment. Measures the tendency of the environment to be reducing (donate electrons) or oxidizing (accept electrons).

Respiration   Energy-yielding metabolism in which oxygen is the terminal electron acceptor for substrate oxidation.

Rhizosphere   The soil that surrounds the plant roots. Active microbial population is higher in this area due to higher nutrients.

Sanitary landfill   A method for disposal of solid wastes in low-lying areas; the deposited wastes are covered with a layer of soil each day.

Secondary sewage treatment   The treatment of the liquid portion of sewage containing dissolved organic matter using microorganisms to degrade the organic matter that is mineralized or converted to removable solids.

Self-purification   Inherent capability of natural waters to cleanse themselves of pollutions based on biogeochemical cycling activities and interpopulation relationships of indigenous microbial populations.

Septic tank   A simple anaerobic treatment system for waste water where residual solids settle to the bottom of the tank and the clarified effluent is distributed over a leaching field.

Sewage treatment   The treatment of sewage to reduce its biological oxygen demand and to inactivate the pathogenic microorganisms present.

Sewage   The refuse liquids or waste matter carried by sewers.

Siderophores   Compounds produced by microorganisms that are involved in the uptake of iron by those microorganisms.

Sink   A reservoir that uptakes a chemical element or compound from another part of its cycle.

Sludge   The solid portion of sewage.

Soil horizon   A layer of soil distinguished from layers above and below by characteristic physical and chemical properties.

Solid waste refuse   Waste material composed of both insert materials— glass, plastic, and metal—and decomposable organic wastes, including paper and kitchen scraps.

Sparge   To pass a gas through a solution.

Spores   (bacterial endospores) A metabolically dormant state of bacteria in which they are more resistant to heat, chemicals, etc.

Surfactant   A natural or synthetic chemical that promotes the wetting, solubilization, and emulsification of various types of organic chemicals.

Symbiosis   An obligatory interactive association between members of two populations, producing a stable condition in which the two organisms live together in close physical proximity, to their mutual advantage.

Synergism   In antibiotic action, when two or more antibiotics are acting together, the production of inhibitory effects on a given organism, which are greater than the additive effects of those antibiotics acting independently; an interactive but nonobligatory association between two populations in which each population benefits.

Tertiary recovery of petroleum The use of biological and chemical means to enhance oil recovery.

Tertiary sewage treatment A sewage treatment process that follows a secondary process, aimed at removing non-biodegradable organic pollutiants and mineral nutrients.

Thermophile Any organism that has an optimum growth temperature above 45°C.

Toxicant An agent that can produce an adverse response (effect) in a biological system, seriously damaging its structure or function or producing death. A toxicant or foreign substance may enter into an environment deliberately or accidentally. Toxicants can be categorized based on source of entry: Non-point source (agricultural run-off, degraded sediment disposal and atmospheric fallout) and Point source (Effluents from manufacturing plants, hazardous waste dispersal sites and municipal waste treatment plants).

Tricking filter system A simple, film-flow aerobic sewage treatment system. The sewage is distributed over a porous bed coated with bacterial growth that mineralizes the dissolved organic nutrients.

Turbidostat A system in which an optical sensing device measures the

turbidity of the culture in a growth vessel and generates an electrical signal that regulates the flow of fresh medium into the vessel and the release of spent medium and cells.

Ultraviolet light (UV) Short-wavelength electromagnetic radiation in the range 100–400 nm.

Vadose zone Unsaturated zone of soil above the ground water, extending from the bottom of the capillary fringe all the way to the soil surface.

Variable nonculturable micro-organism Microorganisms that do not grow in viable culture methods, but which are still metabolically active and capable of causing infections in animals and plants.

Vegetative Cells with an active metabolism.

Vesicular-arbuscular mycorrhizae A common type of mycorrhizae characterized by the formation of vesicles and arbuscules.

Viable Living, or capable of growth.

Visible light Radiation in the wavelength range of 400–800 nm that is required for photosynthesis but can be lethal to some nonphotosynthetic microorganisms.

Volatile organic chemical (VOC) Organic compound that vaporizes into the atmosphere.

Waste waters Under RCRALDR regulations wastes with <1 wt % total organic carbon (TOC) and < 1wt% total suspended solids. All other wastes are considered to be non-waste waters.

Water activity (a_w)   A measure of the amount of reactive water available, equivalent to the relative humidity; the percentage of water saturation of the atmosphere.

Weathering   All physical and chemical changes produced by atmospheric agents.

White rot fungi   Fungi that decompose all components of wood. They produce enzymes that are capable of acting on and biodegrading a wide variety of compounds, including many pollutants.

Wild-type strain   Strain of an organism isolated from nature. The usual or native form of an organism as opposed to a mutant strain.

Windrow method   A slow composting process that requires turning and covering with soil or compost.

Winogradsky column   Glass column with an anaerobic lower zone and an aerobic upper zone, which allows growth of microorganisms under conditions similar to those found in nutrient-rich water and sediment.

Xenobiotic   Compound foreign to biological systems. Often refers to human-made synthetic compounds that are resistant or recalcitrant to biodegradation and decomposition.

Xerophyte   Organism adapted to grow at low water potential, i.e., very dry habitats.

Xerotolerant   An organism able to withstand dryness and capable of growth at low water activity.

Yeasts   A category of fungi defined in terms of morphological and physiological criteria; typically, unicellular, saprophytic organisms that characteristically ferment a range of carbohydrates and in which asexual reproduction occurs by budding.

Zymogenous   A term used to describe opportunistic soil microorganisms that grow rapidly on exogenous substrates.

REFERENCES

Ahmad, W.A. and Zakaria, Z.A. 2004. Bioremediation. In: *Concise encyclopedia of bioresource technology.* (ed.) Ashok Pandey. Food Products Press, New York, pp. 63–70.

Ahmad, W.A., Jaapar, J. and Zahari, M.A.K.M. 2004. Removal of heavy metals from wastewater. In: *Concise encyclopedia of bioresource technology.* (ed.) Ashok Pandey. Food Products Press, New York, pp. 152–157.

Ahmed, N., Badar, U. and Rahan, S. 2004. Resistance and accumulation of heavy metals by indigenous bacteria: Bioremediation. In: *Industrial and environmental biotechnology.* (eds.) Ahmed, N., Foud, M., Qureshi and Khan, O.Y. Horizon Scientific Press, Wymondham UK, pp. 81–102.

Alexander, M. 1999. Bioremediation of metals and inorganic pollutants. In: *Biodegradation and Bioremediation.* Academic Press, New York, pp. 377–390.

Alper, J. 1983. Biotreatment firms rush to market place. *Biotechnology.* 11, pp. 973–975.

Amman, R.I., Ludwing, W. and Schleifer, K.H, 1995. *Microbial Rev.* 59, pp. 143–169.

Amutha, P., Sangeetha, G. and Mahalingam, S. 2001. Dairy effluent induced alterations in the protein, carbohydrate and lipid metabolism of freshwater Teleost fish, *Oreochromis mossambicus. Poll. Res.,* 21, p. 51.

Atkinson, B.W., Bux. and Kasan, H.C. 1998. Waste activated sludge remediation of metal plating effluents. *Water SA,* 24 (4), pp. 355–359.

Atlas, R.M. 1991. *In situ bioremediation.* (ed.) Hinchee, R.E. and Olefenbuttel, R.F. Butterworth Heinemann, pp. 14–32.

Atlas, R.M. and Bartha, R. 2005. *Microbial ecology—Fundamentals and Applications.* Fourth edition, Pearson education (Singapore) Pvt Ltd., Delhi. pp. 511–549.

ATSDR–2001. CERCLA Priority List of Hazardous Substances, (c.f. http://www.atsdr.cdc.gov/clist.html).

Bailey, R.A., Clark, H.M., Ferris, J.P., Krause. S. and Robert L. Strong. 2005. *Petroleum, Hydrocarbons and Coal. Chemistry of the environment.* Academic Press, California, USA. pp. 147–190.

Bedell, G.W. and Darnall, D.W. 1990. Immobilization of nonviable, biosorbent. algal biomass for the recovery of metal ions. In: *Biosorption of heavy metals.* (ed.) Volesky, B. Boca Raton: CRC Press, pp. 313–326.

Berkel R. Van. 1995. Introduction to cleaner production assessments with applications the food processing industry. *UNEP Industry and Environment.* January–March, pp. 8–15.

Beveridge, T. J. and Doyle. R. 1997. Metal ions and bacteria, In: *Microbes and metals.* Mini-review. Ehrlich, H.L. Appl. Microbiol. Biotechnol. 48, p.687.

Binulal, N.S., Zeccharia, A. and Selvin, J. 2003. Bioremediation of oil pollution: present status and future prospects. In: *Biotechnology emerging trends.* (eds.) Selvin. J., Ninawe, A.S., Sugumaran, V.S., Suthumaran, N. and Lipton, A.P. Biotech books, Delhi. pp. 236–243.

Blackburn, J.W. and Hafker, W.R. 1993. The impact of biochemistry, bioavailability and bioactivity on the selection of bioremediation techniques. *Tibtech* 11, pp. 328–333.

Boeniger, M. 1980. The carcinogenicity and metabolism of azo dyes, especially those derived from benzidine. National Institute for Occupational Safety and Health (NIOSH) technical report, U.S. Department of Health and Human Services Washington D.C.

Boopathy, R. 2000. Factors affecting bioremediation technologies. *Biosource Technol.* 74, p. 63.

Bosecker, K. 1993. Biosorption of heavy metals by filamentous fungi, In: *Biohydrometallurgical technologies.* Vol-II,The minerals, metals and materials society, (eds.) Torma, A.E., Apel, M. L. and Brierley, C. L. (Warrendale, Pa), p. 55.

Bosma, T.N.P., Middledrop, P.J.M., Schraa, G. and Zehnder, A.J.B. 1993. Mass transfer limitation of biotransformation: quantifying bioavailability. *Environ. Sci. Technol.* 31, pp. 248–252.

Bouwer, E.J. and Zehnder, A.J.B. 1993. Bioremediation of organic compounds—Putting microbial metabolism at work. *Tibtech* 1, pp. 365–367.

Bumpus, J. A.1995. Microbial degradation of azo dyes, In: *Microbial degradation of health risk compounds.* (ed.) Singh, V. P. Elsevier, Amsterdam. p. 157.

Burd, G. J., Dixon, D. and Glick, B.R. 2000. Plant growth-promoting bacteria that decrease heavy metal toxicity in plants, *Can. J. Microbiol.* 46, p.245.

Burd, G.I., Dixon, G. and Glick, B.R. 1998. A plant growth-promoting *Rhizobacterium* that decreases nickel toxicity in seedlings. *Appl. Environ. Microbiol.* 64, p. 3663.

Cacciatore, D.E. and Mc Neil, M.A. 1995. Principles of soil bioremediation, *Biocycle.* 36, pp. 61–64.

Castello, R.C., Real, J.M.L. and Beck, A.J. 2004. Biodegradation of Polycyclic aromatic hydrocarbons. In: *Concise encyclopedia of bioresource technology.* (ed.) Ashok Pandey. Food products Press, New York. pp 23–29.

Cerniglia, C.E., Gibson, D.T. and Donge. 1994. Metabolism of Benz(a)anthracene by the filamentous fungus *Cunniinghamella elegans. Appl. Environ. Microbiol.* 60, pp. 3931–3938.

Cerniglia, C.E., White, G.L. and R.H. Heflich. 1985. Fungal metabolism and detoxification of polycyclic aromatic hydrocarbons. *Arch. Microbiol.* 143, pp. 105–110.

Chakrabarty, A.M. 1996. Microbial degradation of toxic chemicals: Evolutionary insights and practical consideration. *ASM News.* 62, p. 130.

Chandhry, G. R. and Chapalamadugu. 1991. Biodegradation of halogenated organic compounds. *Microbiological Reviews.* 55, pp. 59–79.

Chatterjee, A.K., Thurn. K.K. and Tyrell, D.J. 1985. Isolation and characterization of Tm5 insertion mutants of *Erwina chrysanthemi* that are deficient in polygluconate catabolic enzymes oligogalactuonate lyase and 3-deoxy-D-glycero-2, 3-hexodiogluconate dehydrogenase. *J. Bacteriol.* 162, pp. 708–714.

Chivukula, M., Sapadars, J.I. and Renganathan, V. 1995. Lignin peroxidase catalysed oxidation of sulfonated azo dyes generates novel sulfophenyl hydroperoxides, *Biochemistry.* 34, p. 7765.

Colwell, E.B. 1990. The ecological effects of oil pollution on littoral communities. Applied Science Publishers, London.

Commission of Engineering and Technical System (CETS), National Research Council. 1993. *In situ* bioremediation, National Academy Press, Washington, DC.

Cookson, J.T.J.R. 1995. *Bioremediation Engineering—Design and Application.* McGraw Hill, New York.

Correia, V.M., Stephenson, T. and Judd, S.J. 1994. Characterization of textile wastewaters—A review. *Environ. Technol.* 15, p. 917.

Cox, M.M. and Battista, J.R. 2005. *Deinococcus radiodurans*—The consummate survivor. *Nature reviews/microbiology.* 3, pp. 882–892.

Czuczwa, J.M. and R.M. Hites. 1984. Environmental fate of combustion-generated polychlorinated dioxins and furans. *Environ. Sci. Technol.* 18, pp. 444–450.

D'Souza, S.F. and Nodkarni, G.B. 1980. Immobilized catalase containing yeast cells preparation and enzymatic properties. *Biotechnol. Bioeng.* 22, p. 2191.

Daly, M.J. *et al.,* 2004. Accumulation of Mn (II) in *Deinococcus radiodurans* facilitates γ-radiation resistance, *Science.* 306, pp. 1025–1028.

Daniel, R. 2004. *Curr. Opin. Biotechnol.* 11:280–285.

Diels, I.Q., Dony. D., Van der Lelie., Baeyens, W. and Mergeay, M. 1995. The CZC operon of *Alcaligens eutrophus* CH34: from resistance mechanism to the removal of heavy metals. *J. Ind. Microbiology.* 14, pp. 142–153.

Dimichele, L. and Malcol, H.Taylor.1978. Histopathological and physiological responses of (*Fundulus hetrochitus*) to naphthalene exposure. *J. Fish Res. Board Can.* 35, p. 1060.

Doraisamy, P. 1994. Anaerobic degradation of aromatic compounds. Ph.D thesis submitted to Tamil Nadu Agricultural University. Coimbatore. p. 180.

Edgehill, R. V. and Finn, R. K. 1983. *Appl. Environ. Microbiology.* 45, pp. 1122–1125.

Efroymson, R. and Alexander, M. 1991. Biodegradation by an Arthobacter species of hydrocarbons partitioned into an organic solvent. *Appl. Environ. Microbiol.* 57, pp. 1441–1447.

Ehlich, H.L. 1997. Microbes and metals. *Appl. Microbiol. Biotechnol.* 48, pp. 687–692.

Ehrlich H. and Brienly, C.L. 1990. *Microbial Mineral Recovery.* McGraw-Hill, New York.

Eisler, R.1985. Mirex hazards to fish, wildlife, and invertebrates: a synoptic review. *U.S. Fish Wildl. Serv. Biol. Rep.* 85:(1.1), p. 42.

Eismanri, F., Becku, P., Kuschk, P.K. and Statmeisters. 1996. Alternative electron acceptors in microbial Coal-conversion wastewater treatment. *Appl. Microbiol. Biotechnol.* 46, p. 604.

EPA. 1980. Ambient water quality criteria for Polynuclear aromatic hydrocarbons. *U.S. Environ. Protection Agency Rep.* 440/5-80-069. pp. 193 –211.

Evans, W.C. and Fuchs. 1988. Anaerobic degradation of aromatic compounds. *Annual Rev. Microbiol.* 42, p. 289.

Feigel, B. J. and Knackmuss, H. J. 1993. Syntrophic interactions during degradation of 4-aminobenzenesulfonic acid by a two species bacterial culture. *Arch. Microbiol.* 159, p. 124.

Ferron, A. and Leggett, W. C.1994. An appraisal of condition measures for marine fish larvae. In: *Advance Marine Biology.* (eds.) Blaxter, J.H.S. Southward, A.J. Academic Press Limited. 30, p. 217.

Flemming, J.T., Sanseverino, J. and Ayler, G.S. 1993. *Environ. Sci.Technol.* 27, pp. 1068–1074.

Fletcher, R.D. 1994. Practical consideration during bioremediation. In: *Remediation of Hazardous waste contaminated soil.* (eds.) Wise, D. and Transtolo, D. Marcel Dekker Inc., New York, pp. 39–54.

Fourest, E. and Roux, J. C. 1992. Heavy metal biosorption by fungal mycelial by-products: mechanisms and influence of pH, *Appl. Microbiol. Biotechnol.* 37, p. 399.

Fuchs, G., Mohamad, M.E.S., Alterschmidt, U., Koch. J., Lack. A., Brackmann, A., Lochmeyer, C. and Oswald, B. 1994. *Biochemistry of aerobic biodegradation of aromatic compounds.* (eds.) Rarledge C. Kluwer Academic Pub, Dordrecht. p. 513.

Gadd, G., Bridge, M.M., Gharieb, J.A. and White, C. 2004. Microbial processes for solubilization or immobilization of metals and metalloids and their potential for environmental bioremediation, In: *Industrial and environmental biotechnology.* (eds).

Ahmed, N., Foud, M., Qureshi and Khan, O.Y. Horizon Scientific Press, Wymondham, UK, pp. 55–80.

Gadd, G.M. 1992. Microbial control of heavy metal pollution. In: *Microbial control of pollution.* (eds.) Fry, J.C., Gadd, G.M., Herbert, R.A., Jones, C.W. and Watson-Graik bridge, I.A. Cambridge University Press, p. 59.

Gadd, G.M. 2000. Bioremedial potential of microbial mechanisms of metal mobilization and immobilization. *Curr. Opi. Biotech.* 11, pp. 271–279.

Gadd, G.M. and White, C. 1993. Microbial treatment of metal pollution-a working biotechnology? *Tibtech.* 11, p. 353.

Gargeda-Cabrera, O.A., Esparza–Garcia,.F. and Pena–Cabriales, J.J. 2003. *Environmental Impact of nitrogen fertilizers in the "Bajio" region of Gnanajuato state, Mexico.* pp. 45–54.

Gheewala, S.H.and Annachatre, A. P. 1997. Biodegradation of aniline. *Water Sci. Technol.* pp. 36–53.

Ghosh, A., Paul..D., Sharma. K., Pandey, P., Prakash, D., Singh, R., Goyal, A., Kaur, H. and Jain,.P.K. 2005. Microbial diversity: Potencial applications in bioremediation. In: *Microbial diversity: Current perspectives and potential applications.* (eds.) Satyanarayana T. and Johri B.N.I.K. International Pvt. Ltd., New Delhi. 505–520.

Greaves, A.J., Phillips, D.A.S. and Taylor, J.A. 1999. Correlation between the biodimination of anionic dyes by an inactivated sewage sludge with molecular structure. Part I: *Literature review.* JSDC. 115, pp. 363.

Hagmar, L., Bellander, T., Andersson, C., Linden, K., Attewell, R. and Moller,T. 1991. Cancer morbidity in nitrate fertilizer workers, International Archives of Occupational and Environmental Health. 63, pp. 63 67.

Harayama, S., Rckik, M., Wassafaller, A. and Bairoch. 1987. *Mol. Gen.Genet.* 210, pp. 241–247.

Hassan, B.,Venkateshwaran, A., Fredrickson,.J.K. and Daly. M.J. 2003. Engineering *Deinococcous geothermalis*—for bioremediation of high temperature radioactive waste environments. *Appl. Environ. Microbiol.* 69, pp. 4575–4582.

Heitzer, A. and Sayler, G.S. 1993. Monitoring efficacy of bioremediation. *Trends Biotechnol.* 11, pp. 334–343.

Helder, T. 1980. Effects of 2,3,7,8-tetrachlorodibenzo-p-dioxin (TCDD) on early life stages of the pike (*Esox lucius* L.). *Sci.Total Environ.* 14, pp. 255–264.

Hernandez. A., Mellado, R.P. and Martinez, J.L. 1998. Metal accumulation and vanadium-induced multidrug resistance by environmental isolates of *Escherichia herdmanni* and *Enterobacter cloacae, Appl. Environ. Microbiol.* 64, p.4317.

Heukelekian, H. 1997. Stream pollution and self purification. In: *Industrial wastes-their disposal and treatment.* (eds.). Rudolfs, W.Allied Scientific Publishers, Bikanr, India.

Hill, M.K. 2004. Persistent, bioaccumulative and toxic pollutants, In: *Understanding Environmental Pollution.* Second edition, Cambridge University Press, UK, pp. 338–349.

Hinchee, R.E., Downey, D.C., Dupont, R.R., Agarwal, P.K. and Miller, R,N. 1991. *J.Hazard. Mater.* 27, pp. 315–325.

Ibanez, J.P. and Umetsu, V. 2002. Potential of protonated alginate beads for heavy metals uptake. *Hydrometallurgy.* 64 (2), p. 89.

Iyer, R.H. 2003. Separation and recovery of radioactive and non-radiactive toxic trace elements from aqueous industrial effluents. *Indian J. Exp. Biol.* 41 (9), pp. 1002–1011.

Jackim, E. and Lake, C. 1978. Polynuclear aromatic hydrocarbons in estuarine and searshore environments. In: *Estuarine interactions.* M.L. Wiley Academic Press, New York, pp. 415–428.

Joerger, T. K., Joerger, R., Olsson, E. and Granqvist, C.G., 2001. Bacteria as workers in the living factory: metal accumulating-bacteria and their potential for material science, *Tibtech.* 19, p. 15.

Jogdand, S.N, 2004. *Environmental Biotechnology.* Himalaya Publishing House, Mumbai.

Joseph, K. and Nagendran, R. 2004. *Essentials of environmental studies.* Pearson Education (Singapore) Pte.Ltd., Delhi.

Jothimani. P., Kalaichelvan, G., Bhaskaran, A., Selvaseelan, D.A. and Ramasamy, K. 2003. Anerobic biodegradation of aromatic compounds. *Indian J. Exp. Biol.* 41 (9), pp. 1046–1067.

Kalaiselvan, G., Banu, K.S.P. and Ramasamy, K. 1997. Lignolytic system and its potential application. *Madras Agri. J.* 84 (11,12), p. 628.

Kamrin, M.A., and P.W. Rodgers.1985. *Dioxins in the environment.* Hemisphere Publ. Corp., New York, p. 328.

Kannaiyan, S. 2001. Bioremediation—A Historical perspective and an overview. In: *Bioremediation of Polluted Soil and Water Ecosystems.* Dep. Environ. Sci. Tamil Nadu Agricultural University, Coimbatore. pp. 1–5.

Kapdan, I.K. and Kargi, F. 2002. Biological decolourization of textile dyestuff containing wastewater by *Coriolus versicolor* in a rotating biological connector. *Enzyme Microb. Techol.* 30, p. 195.

Karna, R.R., Uma, L., Subramanian, G. and Mohan, P.M, 1999. Biosorption of toxic metal ions by alkali extracted biomass of marine Cyanobacterium, *Phormidium valderianum* BDU 30501, *World J. Microbiol. Biotech.* 15, p. 353.

Kaul, S.N. and Szpyrkowicz, C. 1998. Future possibilities of research in anerobic processes. In: *Advances in biotechnology.* (ed.) Pandey, A. Educational Publishers and Distributors, New Delhi, pp. 493–501.

Kaur, A., Sandhu, R.S. and Grover, I.S. 1993. Screening of azo dyes for mutagenicity with Ames/Salmonella assays. *Environ. Mol. Mutagenic.* 22, p. 188.

Keharia, H. and Madamwar, D. 2004. Textile dye and effluent. In: *Concise encyclopedia of Bioresource technology.* (ed.) Pandey A., Food product press, The Haworth Reference Press Inc., New York. pp. 167–175.

Kenaga, E.S. and Goving, C.A.I. 1980. Relationship newtonian water solubility, soil solution and octanol water partition and concentration of chemicals in biota. In: *Aquatic toxicology.* (eds.) Eaton, J.G. Parrish, P.R., Hendrick, A.C., American Society for testing materials.

Khan, I.T. and Jain, V. 1995. Effects of textile industry wastewater on growth and some biochemical parameters of *Triticum aestivum. J. Environ. Poll.* 2, p. 50.

Khan, Z.and Anjaneyulu, Y. 2005. Review on applications of bioremediation methods for decontamination of soils. *Res. J. Chem. Environ.* 9(2), pp. 75–79.

Kim, J.I *et al.* 2002. RecA protein from the extremely radio resistant bacterium *Deinococcous radiodurans:* expression, purification and Characterization. *J. Bacteriol.* 184, pp. 1649–1660.

King, R.B., Long, G.M and Sheldon, J.K. 1992. *Practical Environmental Bioremediation.* Boca Raton FL. Lewis Publishers.

Koarkoutas, Y. and Bant. I.M. 2004. Biosurfactant production and application In: *Concise Encyclopedia of Bioresource technology.* (ed.) Pandey A. Food Product Press. The Haworth Reference Press Inc., New York, pp. 505–515.

Kociba, R.J. and Schwetz, B.A. 1982a. Toxicity of 2,3,7,8-tetrachlorodibenzo-p-dioxin (TCDD). *Drug Metabol. Rev.* 13, pp. 387–406.

Kozdroj, J. and Elsas, J.D.V. 2001. *J. Microbial Methods.* 43:197–212.

Krishnamurthi, G.S.R. 2000. Speciation of heavy metals: An approach for remediation of contaminated soils, In: *Remediation engineering of contaminated soils.* (eds.) Wise D.L., Trantalo D.J., Cichon E.J., Inyang H.I. and Stottmeister U. Marcel Dekker Inc., New York, p. 709.

Krishnaswamy, R. and Wilson, D.B. 2000. Construction and characterization of an *E.coli* strain genetically engineered for Ni (II) bioaccumulation, *Appl. Environ. Microbiol.* 66, p. 5386.

Kulla, H.G. 1981. Aerobic bacterial degradation of azo dyes. In: *Microbial degradation of xenobiotics and recalcitrant compounds.* (eds.) Leisinger, T., Cook, A.M., Hutter, R. and Nuesch. Academic Press Ltd. London, United Kingdom 387.

Lange, C.C., Wackett, L.P. and Daly, M.J. 1998. Engineering a recombinant *D.radiodurans* for organopollutant degradation in radioactive mixed waste environment. *Nat. Biotechnol.* 16, pp. 929–933.

Lee, S.D., and L. Grant. 1981. *Health and ecological assessment of polynuclear aromatic hydrocarbons.* Pathotex Publ., Park Forest South, Illinois. p. 364.

Lena, Q. and Rao, G.N. 1997. Heavy metals in the environment. *J. Environ. Qual.* 26, p. 264.

Lenin-Zaidman, *et al.* 2003. Ring like structure of the *Deinococcous radiodurans* genome: a key to radio resistance. *Science.* 299, pp. 254–256.

Lenski, R.E.1992. Relative fitness: Its estimation and its significance for environmental application of microorganisms. In: *Microbial Ecology: Principles, Methods and Application.* (eds.) Levin, M.A., Seidler, R.J. and Rogul. M. McGraw-Hill, New York, pp. 83–198.

Lichtlen, P. and Schaffner, W. 2001. Putting its fingers on stressful situations: the heavy metal regulatory transcription factor MTF-1, *Bioessays.* 23, p. 1010.

Liu, S. and Suflita, J.M.1993. Ecology and evolution of microbial populations for bioremediation. *Tibtech.* 11 : 8, pp. 344–352.

Lloyd, J.R. and Lovley, D. 2001. Microbial detoxification of metals and radionuclides. *Curr. opi. Biotechnol.* 12, pp. 248–253.

Lorenz, P. Liebeton, K., Niehaus F. and ECK, J. 2002. *Curr. Opin. Biotechnol.* 13:527–577.

Macaskie, L.E., Thomas, A.P. and Lloyd, J.R. 2004. Application of microorganisms to the decontamination of heavy metal bearing waters. In: *Industrial and Environmental Biotechnology.* (eds.) Ahmed, N., Foud, M., Qureshi and Khan, O.Y. Horizon Scientific Press, Wymondham, UK, pp. 103–112.

Mackey, D.M. and Cheery, J.A. 1989. Ground water Contamination: pump-and-treat remediation. *Environ. Sci. Technol.* 23(6), pp. 630–636.

Mahle, N.H., and Whiting, L.F. 1980. The formation of chlorodibenzo-p-dioxins by air oxidation and chlorination of bituminous coal. *Chemosphere.* 9, pp. 693–699.

Manu, B. and Chaudhari. 2000. Anaerobic degradation of stimulated textile wastewater containing azo dyes, *Biores. Technol.* 82, p. 225.

Maria Treja and Quintero. 2003. *Bioremediation of Contaminated soils.* E.B. and CBP Olguin, pp. 179–189.

Masters, G.M. 2004. Treatment of water and wastes. In: *Introduction to Environmental Engineering and Science*. Prentice-Hall of India Private Limited, New Delhi.

Mathewson, Joseph, R. and Grubbs, R.B. 1988. "Innovative Techniques for the Bioremediation of Contaminated Soils," presented at 2nd annual CWPCA Industrial and Hazardous Waste Information Exchange, Oakland, CA.

McMullan, G., Meehan, C., Conneely, A., Kirby, N., Robinson, T., Nigam. P., Banat, I.M., Marchant, R. and Symth, W.F. 2001. Microbial decolourization and degradation of textile dyes. *Appl. Microbiol. Biotechnol.* 56, p. 81.

Mehrotra, S., Sandhir, R., and Chandra, D. 2004. Degradation of Xenobiotics and Bioremediation. In: *Environmental Microbiology and Biotechnology*. (eds.) Singh, D.P. and Dwivedi, S.K. New Age International (P) Limited, New Delhi.

Mergaert, J., Anderon, C., Wouters, A., Swings, J. and Kerster, K. 1992. Biodegradation of polyhydroxy alkanoates. *FEMS Microbial Revs.* 103, pp. 317–322.

Mergeay, M.1991. Towards an understanding of the genetics of bacterial metal resistance. *Tibtech.* 9, pp. 17–24.

Miller, G. Tyler. 2004. *Living in the Environment: Principles, Connections, and Solutions.* 13e. Pacific Grove, CA Brooks/Cole–Thomson Learning.

Moat, A.G. and Foster, J.W. 1998. Metabolism of Aromatic Compounds In: *Microbial Physiology.* 2nd edition, John Wiley and Sons, New York, pp. 166–170.

Moodle, A.D. and Ingledeu, W.J. 1990. Microbial anaerobic respiration. *ADV microbial physical.* 31, p. 225.

Muralisastry, Absar Ahmad, Islam Khan, M and Rajiv Kumar. 2004. Microbial nanoparticles production. In: *Nanobiotechnology.* (eds.) Christ of Niemeyer and Chad Mirkin. Wiley-VCH Verlag GmbH & Co. Kga, Weinherim.

Murugesan, A.G., Deivanayagam, R. and Ruby, J. 2003. Pollution abatement through biopurification of industrial effluents. In: *Biotechnology—Emerging trends.* (ed.) Selvin, J., Ninawae, A.S., Sugunan, V.S., Sugumaran and Lipton, A.P. Biotech Books, New Delhi. pp. 202–211.

Murugesan, K. 2002. Studies on production, purification characterization and crystallization of laccase from a white rot fungus *Pleurotus sajor-caju* and its application in bioremediation of textile dye effluent and dye contaminated soils, Ph.D. Thesis, Madras University.

Murugesan, K. and Kalaichelvan, P.T. 2003. Synthetic dye decolouration by white rot fungi, *IJEB.* 41:9, pp. 1076–1087.

NAS. 1979. Polychlorinated biphenyls. Rep. Comm. Assess. PCBs in Environ., Environ. Stud. Bd., Comm. Nat. Resour., Nat. Res. Coun., Nat. Acad. Sci., Washington DC. p. 182.

National Research Council 1985. *Oil in sea: inputs, facts and effects.* National Academy Press, Washington, DC.

Neely, W.B., *et al.,* 1974. Partition co-efficient to measure bio-concentration potential of organic chemicals in fish. *Environ. Sci.Technol.* 8, p. 113.

Neff, J.M. 1979. *Polycyclic aromatic hydrocarbons in the aquatic environment.* Applied Science Publ. Ltd., London. p. 262.

Nigam, P., Armour, G., Banat, I.M., Singh, D. and Marchant, R. 2000. Physical removal of textile dyes from effluents and solid-state fermentation of dye-adsorbed agricultural residues, *Biores. Technol.* 72, p. 219.

Nony, C.R. and Bowman, M.C. 1980. Trace analysis of potentially carcinogenic metabolites of an azo dye and pigment in hamster and human urine as determined by two chromatographic procedures. *J. Chromotogr. Sci.* 18, p. 64.

Nortemann, B., Baumgarten, J., Rast, H.G. and Knackmuss, H.J. 1986. Bacterial communities degrading amino- and hydroxynaphthalene-2-sulfonates. *Appl. Environ. Microbiol.* 52, p. 1195.

NRCC. 1981. Polychlorinated dibenzo-p-dioxins: criteria for their effects on man and his environment. Natl. Res. Coun. Canada, Publ. NRCC No. 18574. p. 251.

O'morchoe, S.B., Orgnnesitan Sayln, S. and Miller, A.V. 1998. Conjugal transfer of R 68.45 and Fp5 between *Pseudomonas aeruginosa* strains in a freshwater environment. *Appl. Environ. Microbiol.* 5A, pp. 1925–1929.

O'Neill. C, Lopez A., Esteves. S., Hawkes. D.L. and Wilcox, S. 2000. Azo-dye degradation in an anaerobic–aerobic treatment system operating on stimulated textile effluent. *Appl. Microbiol. Biotechnol.* 53, p. 249.

Ochari, E. 1997. *Bioinorganic chemistry: An Introduction.* Align and Bacon, Boston.

Olguin,.E.J. 2003. Cleaner bioprocesses and sustainable development. In: *Environmental biotechnology and cleaner bioprocesses.* (eds.) Olguin, Sanchez, G. and Hernandez, E. Taylor and Francis Limited, London, pp. 1–17.

Olson, J.W., Mehta, N.S. and Maier, R.J. 2001.Requirement of nickel metabolism proteins HypA and HypB for full activity of both hydrogenase and urease in *Helicobacter pylori. Mol. Microbiol.* 39, p. 176.

Pazirandeh, M., Wells, B.M., Ryan, R.L. 1998. Development of bacterial based heavy metal biosorbents: enhanced uptake of cadmium and mercury by *Escherichia coli* expressing a metal binding motiff. *Appl. Environ. Microbiol.* 64, p. 4068.

Pertsova, R.N., Kune, F. and Golovela, I.A. 1984. *Folia. Microbiol.* 29, pp. 242–247.

Prahl, F.G., Crecellus, E. and Carpenter, R. 1984. Polycyclic aromatic hydrocarbons in Washington coastal sediments: an evaluation of atmospheric and riverine outes of introduction. *Environ. Sci. Technol.* 18, pp. 687–693.

Premraj, R. and Doble. M. 2005. Biodegradation of Polymers. *Indian Journal of Biotechnology.* 4, pp. 186–193.

Proulx, D. and De La Novie, J. 1988. Removal of micronutrients from wastewater by immobilized algae, In: *Bioreactor immobilized enzymes and cells: Fundamentals and Applications.* (eds.) Moo-Youn., Elsevier Applied Science, New York, pp. 301–310.

Rajagopalan, R. 2005. *Environmental studies—From crisis to cure.* Oxford University Press, YMCA Library Building, Jai Singh Road, New Delhi, p. 65.

Rajendran, P., Ashokkumar, B., Muthukrishnan, J. and Gunasekaran, P. 2002. Toxicity assessment of nickel using *Aspergillus niger* and its removal from an industrial effluent. *Appl. Biochem. Biotechnol.* 102, p. 201.

Ramel, C. (ed.) 1978. Chlorinated phenoxy acids and their dioxins. *Ecol. Bull.* Stockholm. 27, p. 302.

Rappe, C. 1984. Analysis of polychlorinated dioxins and furans. *Environ. Sci. Technol.* 18, 78A–90A.

Ravera, O. 1989. *Ecological assessment of Environmental Degradation, Pollution and Recovery.* Elsevier Science Publishers.

Reinecke, A.J. and Nash, R.G. 1984. Toxicity of 2,3,7,8-TCDD and short-term bioaccumulation by Earthworms (Oligochaeta). *Soil Biol. Biochem.* 16, pp. 45–49.

Revite, T. and Mavinvurve. 2001. Biodegradable plastics-Bacterial Polyhydroxy alkanoates, *Ind. J. Microbiology.* 41, pp. 233–245.

Rittenhouse, R.C. 1988. Quality—The Critical Element in Liquid Fuel Handling. *Power Engineering,* July.

Roane, T.M. and Pepper, I.I. 2000. Microorganisms and metal pollution. In: *Environmental Microbiology,* (eds.) Maier, R.M., Pepper, I.L. and Gerba, C.B. Academic Press, London, UK, p. 55.

Roberts, J.R., Rodgers D.W., Bailey J.R. and Rorke, M.A. 1978. Polychlorinated biphenyls: Biological criteria for an assessment of their effects on environmental quality. *Nat. Res. Coun.* Canada, Rep. 16077. p. 172.

Rochelle, P.A., Fry, Jc. and Day, M.J. 1989. Factors affecting conjugal transfer of plasmids encoding mercury resistance form pure cultures and mixed natural suspensions of epilithelic bacteria. *J. Gen. Microbiol.* B5, pp. 409–424.

Rogers, J.A., Tedaldi, D.J. and Kavanangh, M.C. 1999. A screening protocol for bioremediation of contaminated soil. *Environ. Progress.* 12(2), pp. 146–156.

Rohrer, T.K., Forney J.C. and Hartig. J.H. 1982. Organochlorine and heavy metal residues in standard fillets of coho and chinook salmon of the Great Lakes-1980. *J. Great Lakes Res.* 8, pp. 623–634.

Rozman, K., Rozman, T. and Greim, H. 1984. Effect of thyroidectomy and thyroxine on 2,3,7,8-tetrachlorodibenzo-p-dioxin (TCDD) induced toxicity. *Toxicol. Appl. Pharmacol.* 72, pp. 372–376.

Safe, S. 1984. Polychlorinated biphenyls (PCBs) and polybrominated biphenyls (PBBs): biochemistry, toxicology, and mechanism of action. *CRC Crit. Rev. Toxicol.* 13, pp. 319–393.

Safe, S.H. 1984. Microbial degradation of polychlorinated biphenyls. In: *Microbial degradation of organic compounds.* (eds.) Gibron D.T., Marcel Dekker, New York, pp. 361–369.

Salkinoja–Salonen, M., Middledrop, P., Brigila, M., Valo. R., Hsggblom, M. and Mcabin, A. 1990. Clean up of old industrial sites, In: *Advances in Applied Biotechnology Series. Biotechnology and Biodegradation.* Vol. 4. (eds.) Kambley, D., Chakrabarthy, A. and Omen. Hopustan, Texas Portifolio Publishing Co. and Gulf Publishing Co., pp. 347–367.

Samuelson, P., Wernerus, H., Svedberg, M. and Stahl, S. 2000. Staphylococcal surface display of metal-binding polyhistydiyl peptides. *Appl. Environ. Microbiol.* 66, p. 1092.

Sand. W., Rohde, K., Sabotke, B. and Zenneck, C. 1992. Evaluation of *Leptospirillum ferrooxidans* for leaching. *Appl. Environ. Microbiol.* 58, p. 85.

Satyanarayanan, U. 2005. Biodegradation and Bioremediation. In: *Biotechnology.* Books and Allied (P) Ltd. Kolkata pp. 718–735.

Sauerback, D.R. and Rietz, E. 1983. Soil chemical evaluation of different extractants for heavy metals in soils, In: Comm. Europe Communities Report EUR 8022. Environ. Eff. Org. Inorg. Contam. Sewage Sludge, CA 99, 193726, p. 147.

Saval, S. 2003. Bioremediation: Clean-up biotechnologies for soils and aquifers, In: *Environmental Biotechnology and Cleaner Bioprocesses.* (eds.) Olguin, E.J, Sanchez, G. and Hernandez, E. Taylor and Francis Limited, Philadelphia, pp. 155–166.

Sayler. G.H., Shields, M.S., Breen, A., Tedford, E.T., Hooper, S., Sirotkin, K.M. and Davis, J.W. 1985. *Appl. Environ. Microbiol.* 48, pp. 1295–1303.

Schmidt, T. and Schlegel, H.G., 1994.Combined nickel-cobalt-cadmium resistances encoded by the *ncc* locus of *Alcaligenes xylosoxidans* 31 *A. J. Bacteriol.* 176, p. 7045.

Schwartz, T.R., R.D. Campbell, D.L. Stalling, R.L. Little, J.D. Petty, J.W. Hogan, and E.M. Kaiser. 1984. Laboratory database for isomer-specific determination of polychlorinated biphenyls. *Anal. Chem.* 56, pp. 1303–1308.

Selvaseelan, D.A. 2001. Bioremediation–A Historical perspective and an overview. In: *Bioremediation of Polluted Soil and Water Ecosystems.* Dep. Environ. Sci. Tamil Nadu Agricultural University, Coimbatore. pp. 18–27.

Sims, R.C., and R. Overcash. 1983 Fate of polynuclear aromatic compounds (PNAs) in soil-plant systems. *Residue Rev.* 88, pp. 1–68.

Singer, M.E., and Finnerty 1984 Microbial metabolism of straight-chain and branched alkanes. In: *Petroleum microbiology.* (ed.). R.M. Atlas. Macmillan, New York, pp. 1–59.

Singh, D.P. and Dwivedi, S.K. 2004. Environmental perspectives of microbiology and biotechnology: an overview. In: *Environmental microbiology and biotechnology.* (eds.) Singh, D.P. and Dwivedi, S.K. New Age International (P) Limited, New Delhi.

Sirota, G.R., and J.F. Uthe. 1981. Polynuclear aromatic hydrocarbon contamination in marine shellfish. In: *Chemical analysis and biological fire: polynuclear aromatic hydrocarbons.* M. Cooke and A.J. Dennis (eds.). Fifth international symposium. Battelle Press, Columbus, Ohio. pp. 329–341.

Sposito, F.G. 2000. The chemistry of soils, In: *Environmental Microbiology.* (eds.) Maier, R.M., Pepper, I.L. and Gerba, C.B. Academic Press, London, UK, p. 406.

Stalling, D.L., Norstrom, R.J. Smith, L.M. and Simon, M. 1985. Patterns of PCDD, PCDF, and PCB contamination in Great Lakes fish and birds and their characterization by principal components analysis. *Chemosphere.* 14, pp. 627–643.

Stein, G. 1987. Oil-Gobbling Bacteria Clean Soil at Site of Park in Carson. *Los Angeles Times.* October 11.

Stolz, A. 2001. Basic and applied aspects in the microbial degradation of azo dyes. *Appl. Microbiol. Biotechnol.* 56, p. 69.

Stolzenburg, T. and Sullivan, J. 1983. *Dioxin: a cause for concern?* Publ. WIS-SG-83-141. Univ. Wisconsin, Madison. p. 21.

Subbo Rao, N.S. 2000 Biodegradation of pesticides and Pollutants. In: *Soil Microbiology.* Oxford and IBH Publishing Co. Pvt. Ltd., New Delhi, pp. 303–325.

Suess, M.J. 1976. The environmental load and cycle of polycyclic aromatic hydrocarbons. *Sci. Total-Environ.* 6, pp. 239–250.

Swain, W.R. 1983. An overview of the scientific basis for concern with polychlorinated biphenyls in the Great Lakes. In: *PCBs: Human and Environmental Hazards.* (eds.) D'Itri, F.M. and Kamrin, M.A. Butterworth Publ. Woburn, M app. pp. 11–48.

Taghavi, S., Delanghe, H., Lodewyckx, C., Mergeay, M. and Lellie, D.V. 2001. Nickel-resistance-based minitransposons: new tools for genetic manipulation of environmental bacteria. *Appl. Environ. Microbiol.* 67, p. 1015.

Tan, I.K.P. 2004. Polyhydroxy alkanoate production from renewable resources. pp. 653–662.

Tanaka, M. *et al.,* 2004. *Deinococcous radiodurans* transcriptional response to ionizing radiation and desiccation reveals novel proteins that contribute to extreme radio resistance. *Genetics.* 168, pp. 21–33.

Troy, M.A.1994. Bioengineering of soils and ground waters In: *Bioremediation.* (eds.) Banker, K.H. and Herson, D.S. NewYork, McGraw-Hill Inc., pp. 173–220.

Tyagi, P., Buddhi, D., Chodhary, R. and Sawheny, R.L. 1999. Degradation of groundwater quality due to heavy metals in industrial areas of India-A review. *IJEP.* 20, pp. 174.

Umminger, B.L. 1970. Physiological studies on super cooled kill fish *Fundulus leteroelitus* (III) Carbohydrate metabolism and survival at subzero temperature. *J. Exp. Zool.* 173, p. 159.

UNEP. 1996. Sustainable production and consumption Industry and Environment, 19(3), pp. 4–5.

Van Noort, P.C.M. and Wondergem, E. 1985. Scavenging of airborne polycyclic aromatic hydrocarbons by rain. *Environ. Sci. Technol.* 19, pp. 1044–1048.

Vander Meer, J.R., de Vos, W.M., Harayama, S. and Zehnder, A.J.B. 1992. *Microbial Rev.* 56, pp. 677–694.

Vasudevan, P., Padmavathy, V., Tewari, N. and Dhingra, S.C. 2001. Biosorption of heavy metal ions. *Journal of Scientific and Industrial Research.* 60, pp. 112–120.

Vos, J.G. 1978. 2,3,7,8-tetrachlorodibenzo-para-dioxin: effects and mechanisms. In: Chlorinated phenoxy acids and their dioxins. (eds.) Ramel, C. *Ecol. Bull.* (Stockholm) 27. pp. 165–176.

Wagner, K., Bayer, K., Claff, R., Evans, M., Henry, S. and Hodge, V. 1986. *Remediation action technology for waste disposal sites.* 2nd edition, ParkRidge, New Jersey, Noyes, Data Corporation.

Walia, S.K. and Khan, A.A. 1992. Molecular basis for bacterial degradation of Polychlorinated biphenyls. In: *Industrial biotechnology.* (eds.) Malik, V.S. and Srridar, P. Oxford and IBH Publishing Co. Pvt. Ltd., New Delhi, pp. 415–419.

Wenzel, W.W. 1999. Manipulating rhizosphere chemistry to control metal and organic contaminant availability and implications to Phytoremediation. In: 2nd International conference on contaminants in soil environment in the Australia-Pacific region, New Delhi.

White. *et al.* (1999). Genome sequence of the radio-resistant bacterium *Deinococcous radiodurans*-RI. *Science.* 286, pp. 1571–1577.

World Commission on Environment and Development (WCED). 1987. *Our common future.* Oxford: Oxford University Press.

Yan Clemput, O. and Hera, C.H. 1994. Fertilizer nitrogen use and efficiency in different cropping systems presentation at the XV Congreno International de la Ciencia del Suelo, Acapulco, Mexico, July.

Young Land Yu, J. 1997. Ligninase-catalysed decolourization of synthetic dyes. *Water Res.* 3. p. 1187.

Zhargyand Miller, R.M. 1992. Enhanced Octaecane dispersion and biodegradation by a *Pseudomonas* rhamnolipid surfactant (biosurfactant). *Appl. Env. Microbiol.* 58, pp. 3276–3282.

Zimmerman, T., Kulla, H.G. and Leisinger, T. 1982. Properties of purified orange II azo reductase the enzyme initiating azo dye degradation by *Pseudomonas* KF44 EM. *J. Biochem.* 129, p. 197.

Zollinger, H. 1987. *Colour chemistry–synthesis, properties and applications of organic dyes and pigments.* VCH, New York, p. 92.

INDEX

9 798224 165377